SHIPS AND NAVAL ARCHITECTURE

SHIPS AND NAVAL ARCHITECTURE

(S.I. UNITS)

by

R. MUNRO-SMITH, M.Sc., C.Eng., F.R.I.N.A.

*Formerly Senior Lecturer in Naval Architecture
and Shipbuilding
at the University of Newcastle upon Tyne*

*Formerly Lecturer in Naval Architecture
at the University of Liverpool*

THE INSTITUTE OF MARINE ENGINEERS

TO
ALL THOSE
WHO SET OUT IN SHIPS
IN THE PURSUIT OF
THEIR CALLING

Copyright © First published 1973
SHIPS AND NAVAL ARCHITECTURE

This book is copyright under Berne Convention. All rights reserved. Apart from any fair dealing for the purpose of private study, research, criticism or review—as permitted under the Copyright Act 1956, no part of this publication may be reproduced, stored in a retrieval system or transmitted in any form or by any means, electronic, chemical, mechanical, optical photocopying recording or otherwise, without the prior permission of the copyright owners. Enquiries should be addressed to The Institute of Marine Engineers, 76 Mark Lane, London EC3R 7JN.

ISBN 0 900976 03 2

Printed by Hobbs the printers Ltd, Southampton, SO9 2UZ.

PREFACE

Shipbuilding, although old and traditional is an art which never stands still. Consequently, new trades, new materials and new techniques are always being introduced. However, there has never been with ships a new machine or method of construction, which, on its introduction rendered everything else obsolete. Change in ships must be slow because of the cost of change. The keystone of any proposed change is the cardinal requirement—reliability.

Technological progress in shipping in recent years has surpassed anything which has gone before. This progress is evident in every branch of the shipping industry of this country and includes automation, cargo handling techniques, work study, ship design, propulsion methods and operational techniques.

In spite of the developments which have taken place in all forms of transport, especially air transport, the sea remains the most important connecting link between the nations. The ship is for man an important vehicle of trade; its design and construction a major industry; its operation a work which calls for many skills and team-work.

This book has been written primarily to meet the needs of engineers and navigators serving in the Merchant Navy. It is hoped that it will be found useful and interesting to those in Shipbuilding and Shipping as well as to the legion of people interested in the sea and ships.

A ship at sea is an entity of its own, a small community where law and order, safe conduct, daily routine and domestic and social services all have to be effected as efficiently, if not more so, as on land, and there must exist on board a strong central authority. Such authority is vested in the hands of one person—the master.

The family of ships is a fascinating one and a vigorous one; many people are employed in creating it, ministering to its needs, and keeping it in action.

R. Munro-Smith

CONTENTS

Part I

The Sea and Ships

Chapter		Page
1	The Sea	1
2	Seaworthiness	7
3	Ships	11
4	Merchant Ship Types	23
5	The Ship Girder and Structural Details	43
6	Statutory Regulations	77
7	I.M.C.O.	101

Part II

Naval Architecture

Chapter		Page
8	Glossary of Terms	111
9	S.I. Units	119
10	Calculation of Areas, Moments etc	127
11	Transverse Stability	165
12	Trim	189
13	Watertight Subdivision	205
14	Strength of Ships	217
15	Vibration of Ships	257
16	Rudders and Oscillations	265
17	Resistance and Powering	285
18	Propulsion and Propellers	301

Index

PART I

THE SEA AND SHIPS

It can, indeed, be said that the face of the sea is always changing. The surface waters move with the tides, stir with the wind and rise and fall to the endless forms of the waves.

Man has built ships of many types to venture out on its surface and later found means of descending to the shallow part of its floor. The ocean is the earth's greatest storehouse of minerals. During long years the sea has joined men and in some instances land has separated them.

The world is a planet dominated by its great mantle of ocean in which continents are intrusions of land above the all-encircling sea.

CHAPTER 1
The Sea

Salinity; Temperature; Density; Pressure; Light; Tides; Waves; Beaufort Scale.

It is desirable, indeed essential, that the ship designer and the personnel directly engaged in the operation of ships should know something about the environmental conditions in which a ship operates.

These may be divided into two categories:
a) the environment external to the ship which affects the ship as a whole and its exposed equipment. These conditions are caused by the sea and weather conditions.
b) the internal environment which affects personnel and internal equipment. These conditions are temperature, humidity, ship motions, vibration, noise etc.

The Marine Branch of the Meteorological Office encourages all mariners to take an interest in meteorology. The seaman is so dependent on the weather, not only for the safety of his ship but also for his personal comfort that an interest in meteorology on his part is vital.

The Sea

The sea is the great mother of life.

The oceans and seas of the world cover an area of about 360,000,000 square km or about two-thirds of the earth's surface. They occupy a volume of $1,350 \times 10^6$ cubic km, fifteen times greater than that of the land above sea level and fill a vast irregular depressed basin with an average depth of 3,850 metres, more than five times the average height of the land which in spite of the great mountain systems is about 700 metres.

The greatest depth of the ocean, 10,600 metres, off the coast of Japan, is not much more than the greatest height of the land, 8,900 metres. Less than 2 per cent of the sea floor lies at depths greater than 5,500 metres.

In spite of the vast volume of the water of the oceans and seas it should be remembered that if the world is reduced to the size of a large orange the ocean becomes a mere wetness on its skin.

Water is an almost universal solvent. Sea water contains in very dilute solution all the elements which compose the minerals of the earth's crust. It has been calculated that there are enough salts in the sea to cover the entire globe, when dried, with a layer 45 m thick or the land area with a layer nearly 150 m thick. In this solution common salt—sodium chloride—is certainly the most important constituent.

The chemical composition of sea water may be listed as follows:

	per cent		per cent
Sodium	30·4	Sulphate	7·7
Magnesium	3·7	Bromine	0·19
Calcium	1·2	Carbonate	
Potassium	1·1	and	0·35
Strontium	0·04	Bicarbonate	
Chlorine	55·2	Boric Acid	0·07

Salinity

The total salt content is called salinity. Salinity is directly related to density and to temperature. Certain seas are more salty than others. The North Atlantic has an average salinity of 35 parts per thousand but the Red Sea, where evaporation is intense is more concentrated and has a salinity of 40 per thousand. The water of the Baltic is very dilute owing to the melting of ice and the inflow of many large rivers so that its salinity is about 7 per thousand.

Temperature

In the sea the changes of temperature which occur from place to place and season to season are small compared with those that take place on land. The height of summer and the depth of winter never change the surface temperature of the open sea anywhere more than 10°C and generally the seasonal difference is much smaller. The hottest seas in the world are the Red Sea and the Gulf of Oman where temperatures of over 30°C are common; the coldest seas are in the Antarctic where temperatures of $-1\cdot5°C$ are usual in the higher latitudes. This range is small compared with what takes place on land where the temperature may vary from many degrees below freezing in the Antarctic to 65°C in the deserts of central Asia and where a change of 10°C from day to night is quite common.

Density

The density of sea water depends on the salinity and on the temperature, and varies slightly from place to place and at different seasons of the year. The variation of surface salinity in the seas and oceans is, as stated above, primarily dependent on the difference between the average evaporation and the average rainfall.

The mean value normally assumed for calculations in naval

architecture is 1·025 which is 1,025 kg per cubic metre or 0·975 m³ to the tonne.

Pressure

At sea level the pressure of the atmosphere is 10·14 N per cm². Below the surface of the sea the pressure increases by about one atmosphere every 10 metres depth so that at the bottom of the ocean basins the pressure is enormous. Since, however, water is almost incompressible and since this pressure bears in all directions both inside and outside the cavities and living tissues of the body, even the most fragile organisms are unaffected. In the same way a metal box, sunk to the bottom of the sea, keeps its shape even at the greatest depths if the sides are perforated so that the pressure of the water can bear both inside and out. If there are no holes the box is squashed flat long before it reaches the bottom. Sunken ships, thus, remain unchanged in shape except for any tanks or compartments into which the water cannot easily penetrate; they will burst inwards.

Light

In the sea light becomes altered both qualitatively and quantitatively since 10 per cent or more of the light reaching the surface is lost by reflection and in passing downward through the water its intensity and spectral composition are modified. Even pure water absorbs light rapidly compared with air, and the yellow, orange and red components are absorbed more quickly than the blue, green and violet.

At about 250 metres depth the light intensity is reduced to 0·001 per cent of the surface value; at 500 metres the limit for vision of fish is reached and below 1,000 metres there is no perceptible light from the surface.

Tides

All heavenly bodies exert some attractive force on the earth but the influence of all except the sun and moon is unimportant because of their distance from the earth. The moon exerts about twice the pull of the sun since it is so much nearer to the earth. The amount of attraction varies inversely as the square of the distance, and the moon, therefore, attracts the part of the earth, which is nearest to it more strongly than the parts which are farthest away.

When the earth, sun and moon are in line their combined pull produces the largest movement of the water, the spring tides. When the three bodies are at right angles this gives the smallest movement, the neap tides. These happen at fortnightly intervals.

The earth rotates on its axis once in twenty-four hours and each meridian in turn comes opposite the moon, so that in each rotation of twenty-four hours there are two high and two low tides with roughly six hours between a high tide and the next low.

Waves

The water of the ocean is never still. It is blown into waves by the wind; it rises and falls with the tides and in many places there are definite currents either permanently in one direction or changing with the tide or with the season. The waves caused by the action of the wind on the surface of the sea, are the most familiar and at times the most spectacular of all the motions of the oceans and the seas. The effect of the wind varies from tiny ripples on a pond to the mighty rollers of the North Atlantic. The precise mechanism by which the transfer of energy takes place is not known.

A wave is an oscillation of the water particles at the surface but the water itself is not involved in the forward movement which one sees when watching the wave. A cork floating on a pond in which there is no current simply bobs up and down upon the waves but unless it is pushed along by the wind itself, it changes its position only very slowly. Deep sea waves give the impression of a mass of water moving bodily forward. In deep water, this is only an optical illusion. Small floating objects will be found to have no sustained forward motion, and merely rise or fall with the passage of the wave beneath.

Waves generated by local winds are generally referred to as SEA and those which have travelled out of their area of generation are termed SWELL. Sea waves have relatively peaky crests while swell waves are generally lower with more rounded tops.

The following definitions are used in describing a wave:

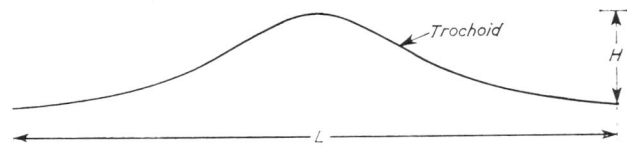

FIG. 1-1—*Ocean wave form.*

a) *Speed (V)* Usually expressed in knots is the speed at which individual waves travel.

b) *Length (L)* is the horizontal distance between successive crests or successive troughs.

c) *Period (T)* expressed in seconds is the time interval required for the passage of successive crests or successive troughs past a given point.

d) *Height (H)* is the vertical distance between the top of a crest and the bottom of a trough.

Mathematical investigation of the motion of waves in water classifies the waves in three types:

1) capillary waves or ripples
2) translation waves of which the tidal wave is an example
3) oscillating waves sometimes called deep-water waves.

The ship designer is, in the main, concerned with the third type. Consideration of the motion of water when disturbed into waves is of importance in the study of resistance, rolling and strength.

The commonly accepted theory of waves in water was originally due to Franz Gerstner in 1804 and developed much later by Rankine and Froude. This "trochoidal theory" as it is termed has been adopted for certain standard calculations such as that for longitudinal strength. A trochoidal curve is the path traced out by a point on the radius of a rolling circle. This is dealt with in more detail in Chapter 14.

Wind

The main influence of the wind is felt indirectly through the waves it generates on the sea surface. The severity of the waves depends on the velocity of the wind, its duration and the distance over which it acts.

Wind force is expressed numerically on a scale generally from 0 to 12. The scale was devised by Captain, afterwards Admiral, Sir Francis Beaufort in the year 1808 for use on board ships of the Royal Navy. Since Beaufort's time many changes have taken place and later in 1941 the International Meteorological Organization agreed to the use of a sea criterion by which the wind force was judged from the appearance of the sea surface. This specification is given in Table 1.1.

Table 1.1. The Beaufort Scale

Beaufort Scale Number	Descriptive Terms	Limits of Wind Speed in Knots
0	Calm	Less than 1
1	Light Air	1–3
2	Light Breeze	4–6
3	Gentle Breeze	7–10
4	Moderate Breeze	11–16
5	Fresh Breeze	17–21
6	Strong Breeze	22–27
7	Near Gale	28–33
8	Gale	34–40
9	Strong Gale	41–47
10	Storm	48–55
11	Violent Storm	56–63
12	Hurricane	64 and over

Wind speed measured at a standard height of 10 metres above sea level.

CHAPTER 2

Seaworthiness

Seaworthiness is a quality that is profoundly affected by many factors. From the point of view of owners and underwriters a degree of uniformity of standard of strength for maximum displacement is considered necessary as a warranty for seaworthiness. In this uniformity of practice the classification societies are specially competent to control.

A ship may be said to be seaworthy when it is in all respects fit to carry the cargo in good condition—as far as protection from the sea is concerned—and deliver it at the port of destination. This is a simple definition and the term "seaworthiness" is a very comprehensive one, embracing the main elements of ship design, and ship operation in strength, stability, freeboard, machinery, steering gear, etc. If it be assumed that the propelling machinery is in good order and that the steering gear, equipment, etc., are entirely satisfactory, then the important elements are: Strength; Freeboard and Stability.

Strength

Obviously, sufficiency of strength is one very important quality necessary for the safety of a ship on a sea voyage. If the structure is not robust enough for the seas it will meet and the loads impressed upon it, it will fail in its primary function. The thickness of the steel plating and the dimensions of the frames, beams, girders, etc.—generally called the scantlings—of the main structure necessary to give the required strength to the ship are prescribed by the rules of the classification societies. These societies are authorized under the Merchant Shipping Acts to assign to a vessel a maximum permissible draught in association with their rules. If it is proposed to use for a ship of specified dimensions scantlings less than prescribed in the rules, then the maximum draught would have to be decreased in order to reduce the forces exerted by the sea on the hull.

If, on the other hand, the scantlings of the structure are in excess of that complying strictly with the societies' rules the maximum draught cannot be exceeded, since no increase of strength will compensate for the deficiency in seaworthiness that would follow any further increase of draught. Consequently the maximum permissible draught is determined by considerations in addition to that of strength.

Freeboard

There is ample evidence, derived from many years of experience, to establish that there is a minimum height of side of a ship above still water level which will be sufficient to ensure:
 a) the satisfactory navigation and working of the ship;
 b) adequate protection for the cargo hatches and ship's appliances from damage by the sea;
and c) the cargo being delivered in good condition.

This minimum height of side is included in the term "freeboard". The Load Line Convention defines freeboard thus: "the freeboard assigned is the distance measured vertically downwards amidships from the upper edge of the deck line to the upper edge of the related load line".

The deck to which the freeboard is measured is normally the uppermost complete deck exposed to weather and sea, which has permanent means of closing all openings in the weather part and below which all openings in the sides of the ship are fitted with permanent means of watertight closing. In a ship having a discontinuous freeboard deck, the lowest line of the exposed deck and the continuation of that line parallel to the upper part of the deck is taken as the freeboard deck.

Freeboard, together with the maximum permissible draught, makes up the total depth of the ship to the deck to which the freeboard is measured. Any increase or decrease of draught gives rise to a corresponding decrease or increase of freeboard respectively.

Minimum freeboard in association with a structure of ample strength may be different for different voyages or for the same voyage at different seasons of the year in order to make allowance for the weather conditions likely to be experienced. Ships not more than 100 metres in length which enter any part of the North Atlantic during the winter season require more freeboard. Ships require more freeboard in all waters during the winter season than in summer.

Vessels navigating fresh water are permitted so long as they are in fresh water to have a reduced freeboard. The reduction amounts broadly to the sinkage in passing from sea to fresh water and is about 19 cm for a ship of length 170 metres. The prime purpose in assigning a freeboard for fresh water to an ocean going ship is to make it possible to load in fresh water to a draught which will be suitable on its arrival in the sea.

Reserve of Buoyancy

The watertight volume of a ship above the waterline is called the reserve of buoyancy. It can be defined as the buoyancy a ship can call upon to meet losses of buoyancy due to damage of the main hull. The actual reserve at any moment depends upon the draught of water, but is generally used only in reference to the load draught.

Reserve buoyancy is not provided specifically to compensate for any losses of buoyancy that may be incurred; its use in the general working of the ship is to provide a sufficiency of freeboard to make the vessel seaworthy. In passenger ships water-tight subdivision is provided so that a ship may be able to withstand serious damage to the main structure with resulting loss of buoyancy and continue to float with certain prescribed extents of damage. It is not the prime function of reserve buoyancy to meet leakages caused by extensive damage.

Large reserve of buoyancy is not basically a governing factor in the assessment of freeboard for cargo ships since it is possible to make the ship seaworthy with the water excluding fittings required by the Load Line Convention.

Deck Erections, Superstructures and Sheer

A flush deck ship is one which has no superstructure on the freeboard deck. Assuming a minimum freeboard being prescribed for a flush-deck ship it is clear that the addition of a deck-house extending across the ship will certainly make the vessel to some extent more seaworthy. With a structure of ample strength it would be permissible to reduce the freeboard. The extent to which an erection can contribute to general seaworthiness depends upon its position in the ship and whether it is watertight or only weather-tight. Credit in the form of deductions from freeboard depends upon the extent and effectiveness of the erection.

Sheer and deck erections contribute to the seaworthiness of the ship and consequently reductions in freeboard can be made. However, the usefulness of sheer is limited since sheer would require to be very large to prevent seas coming over near amidships.

Stability

The International Load Line Convention assumes that the nature and stowage of the cargo, ballast, etc., are such as to secure sufficient stability of the ship and the avoidance of excessive structural stress.

Responsibility for loading the ship to give adequate stability for seaworthiness is placed upon the master who is given the authority to do what is necessary for the proper stowage of the cargo to secure that measure of stability.

The principal factor in deciding any ship's stability is the manner of distribution of the cargo in the vertical direction. Any cargo ship can secure the desired stability by proper loading although this may in some instances involve the expensive expedient of ballasting.

The Load Line Rules state:
a) The master of every new ship shall be supplied with sufficient information, in an approved form, to enable him to arrange for the loading and ballasting of his ship in such a way as to avoid the creation of any unacceptable stresses in the ship's structure.

b) The master of every new ship which is not already provided with stability information under an international convention for the safety of life at sea in force shall be supplied with sufficient information in an approved form to give him guidance as to the stability of the ship under varying conditions of service.

No definite standard of stability is laid down except for the damaged condition. The stability is to be sufficient to enable the ship to withstand the final stage of flooding a compartment or compartments within the floodable length. So far it has not been possible to devise a generally acceptable criterion for a minimum measure of stability either in terms of metacentric height or of righting moment. Any such measure is dependent upon many factors difficult to define so as to administer an Act of Parliament.

The desirable amount depends somewhat on the type of vessel but in very broad terms lies in a metacentric height of between 0·4 m and 0·75 m.

The control of stability is the concern of the master and he must ensure that the ship is stable and that it does not become unstable. It is commonly stated that a ship will have the most comfortable motion at sea when the metacentric height is small. In general this is true for large ocean-going ships in the seas they encounter but it must be recognized that this easy motion obtained by the small metacentric height will leave a very small margin upon which to draw in an emergency.

Stability is an essential, indeed a vital, element in a ship's seaworthiness and an easy motion must not be obtained at any sacrifice of safety.

One of the main objects in loading a ship is to so distribute the cargo that satisfactory stability and trim are secured without the use of ballast. This then makes the entire ballast capacity of the ship available for the adjustment during the voyage of draught, trim and metacentric height or for use in an emergency.

CHAPTER 3
Ships

Primary Duty; Statistics; Propulsion and Fuel Analysis; Ship Classes; The Family Tree of Ships; Development; Requirements of Hull and Machinery; Marine Nuclear Power.

The many different kinds of cargoes carried and the economic pressure upon shipowners in a highly competitive system to earn freights with least expenditure have resulted in the development of particular types of ships for specific purposes.

The primary duty of a merchant ship is to convey cargo or passengers from one port to another where land transport is either uneconomical or impracticable. Air transport is now a very important factor in the conveyance of passengers and freight all over the world.

In spite of the developments which have taken place in all forms of transport, especially air transport, the sea remains the most important connecting link between the nations. The ship is for man an important vehicle of trade; its design and construction a major industry; its operation a work which calls for many skills and team work. The carriage of goods by sea has been little affected by other transport activities as is shown by the 247 million plus tons gross of ships afloat in the world today, a total which is made up of over 55,000 individual vessels, literally of all sorts, sizes and shapes.

Before giving consideration to how the ship family as an international unit is composed it is of interest to have some idea of the proportions applicable to the principal seafaring nations. The attitude of any nation to the problem of sea transport is governed by internal conditions.

Table 3.1 gives some statistics for several countries for the years 1939 and 1971.

The ever-growing demand for oil and its derivatives has had its effect on the world tonnage of tankers. This is shown in Table 3.2.

Ore and Bulk Carriers
The world total of ore and bulk carriers has increased enormously and now represents 21·8 per cent of all steamships and motorships.

The propulsion analysis and fuel analysis of the world merchant fleet tonnage are given in Table 3.3.

Table 3.1.

Year	WORLD		Country	1939		1971	
	Gross Tonnage	No. of Ships		No. of Ships	% World Gross Tonnage	No. of Ships	% World Gross Tonnage
1939	68·5 m	29763	United Kingdom	6,722	26·2	3,785	11·1
1971	247·2 m	55041	U.S.A.	2,853	16·6	3,327	6·6
			Liberia	—	—	2,060	15·6
			Japan	2,337	8·2	8,851	12·3
			Norway	1,987	7·0	2,814	8·6
			Germany (West)	2,459	6·6	2,826	3·5
			Italy	1,227	5·0	1,690	3·3
			France	1,231	4·3	1,399	2·8
			Greece	607	2·6	2,056	5·3
			Russia (U.S.S.R.)	699	1·9	6,575	6·6

SHIPS

Table 3.2.—World Tonnage of Tankers

Year	World Merchant Fleet Gross Tonnage	STEAM % World Gross	MOTOR % World Gross	Total STEAM and MOTOR % World Gross
1939	68·5 m	8·0	8·9	16·9
1971	247·2 m	22·8	16·1	38·9

Table 3.3.

Year	STEAMSHIPS % World Gross	MOTORSHIPS % World Gross	FUEL Coal %	FUEL Oil %
1939	75·4	24·6	45	55
1971	35·4	64·6	2 approx.	98

The figures given in the above tables have been taken from "Lloyd's Register of Shipping Statistical Tables". This is a most informative publication giving statistical details back to the year 1904.

It will be seen from Table 3.1 that Japan has the largest number of ships and Liberia has greatest gross tonnage. The total fleet of the British Commonwealth consists of 7,409 individual ships of total 38·6 m gross tons.

At the beginning of the present century the British Mercantile Marine comprised about 50 per cent of the total world tonnage. Today as the Table shows it is in comparison relatively small. The reasons for this decline have been given in part as (a) the development of fleets under flags of convenience, (b) the maintenance and development of national fleets and (c) the direction of cargo by governments to their national shipping, i.e. flag discrimination.

Merchant ships can as a first analysis be divided into two classes; cargo ships and passenger ships.

Cargo Ships

In this class the provision of adequate space to carry the cargo and facilities for handling the cargo are two very important factors. Figure 3-1 shows an outline profile of a typical cargo ship having two decks. Formerly in this type of ship the propelling machinery was placed amidships with the cargo holds situated forward and aft of this space, the spaces between the decks—the 'tween decks—being also available for the stowage of cargo. Now the tendency is to locate the machinery aft as indicated in Figure 3-1. The use of the amidship portion of the hull for carrying cargo instead of the propelling unit is a definite improvement but it also raises the problem of trim. Fortunately this can usually be solved by having a midship deep tank which can be available for cargo as well as water ballast. A popular compromise between the ship with propelling machinery amidships and that with the machinery aft is to locate the propelling unit so that there are, say, three holds forward and one hold aft of the machinery space. This arrangement eases the problem of trim and strength.

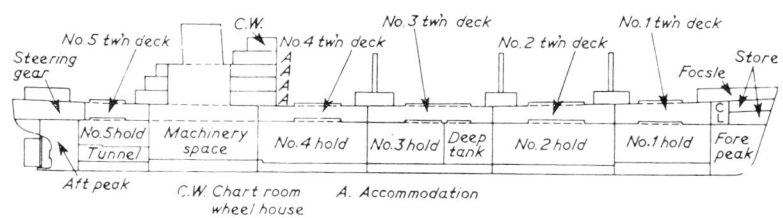

Fig. 3-1—*Profile of typical 2-deck cargo ship.*

Access to the cargo spaces is through hatchways, large openings in the deck, and these must have portable coverings which can be readily removed when the ship is in port but must also be weathertight when the ship is at sea. Each hatch must have provision for loading and unloading cargo and this consists of derricks or cranes. The derricks are fitted either to the masts or to special derrick posts. The derricks are operated by means of steam or electric winches. The use of the latter is now standard cargo liner practice. The steam winch has a relatively low first cost but high running cost and is gradually being replaced by the electric winch.

The cargo ship is not always operating fully loaded and since the ship in the light condition would present a considerable area exposed to the wind and as the propeller in this condition would be only partly immersed it is essential to provide space for the carriage of water ballast. Such spaces are the fore and after peaks and portions of the double bottom tanks. Double bottom tanks are also used for oil fuel and boiler feed water.

Passenger Ships

In the passenger ship some of the features provided in the cargo ship also apply with the obvious addition of accommodation for the passengers. This involves the provision of cabins and public rooms such as dining rooms, lounges, smoke rooms, theatre, swimming pool, promenade space etc. The controlling factor is space for the number of passengers it is intended to carry. This generally means more decks to accommodate the passengers particularly where an appreciable number of passengers are involved. These additional decks are usually fitted above the weather deck and create the characteristic feature of the passenger ship—the superstructure above the uppermost continuous deck. The fitting of extensive superstructures creates its own problems as will be shown later.

The comfort of passengers is most important and has received a great deal of attention including air conditioning and the fitting of anti-rolling fins. Excessive pitching has still to be eliminated and, no doubt, this will be achieved in time.

Ship Types

The family tree of ships is shown in Table 3.4. There are many ways of classifying ship types and many sections into which it is possible to place any particular kind. As far as ocean-going ships are concerned however they either carry passengers with a small amount of cargo, or cargo without any reference to passengers or passengers and cargo together. There are in service large passenger ships in which cargo space has been eliminated, the machinery placed right aft and the entire internal arrangement devoted to the best possible layout of cabins and public rooms.

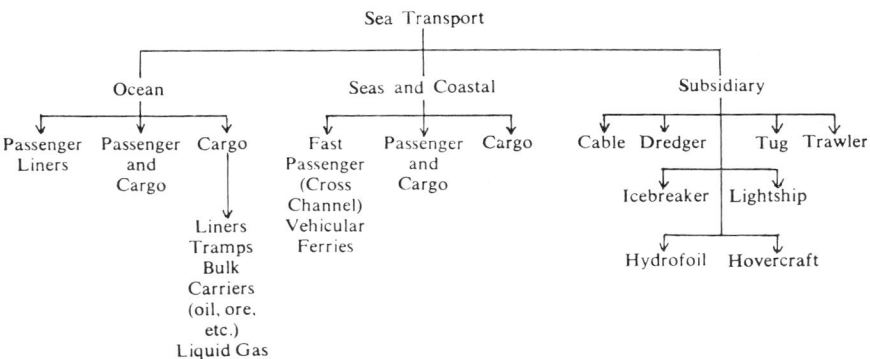

Table 3.4.—The Family Tree of Ships

The Atlantic has produced a unique class of ship, the mammoth Queens and the like; ships which many people consider are too expensive to run and serve only to further national prestige; ships which, above all others are subject to the ever increasing pressure of air transport. After the mammoths there are the more rational types of passenger ship, smaller, less expensive and not so fast but nevertheless quite luxurious. Such a vessel is the cross-channel ship. Ships of this type have high speeds for their size, 22 knots and above and are essentially passenger ships, little or no cargo being carried. Associated with the cross-channel ships are the passenger-vehicular ferries in which the arrangement below deck is subordinated to garage space. On the upper decks passenger accommodation is provided. These vessels can operate as a drive-on, drive off type of ship and may have hydraulically operated doors at the bow and the stern.

Cargo Ships

Cargo ships are plain, utilitarian craft as the name indicates but with excellent accommodation for the crew. Cargo carriers may be put into three main divisions:

1) General trading
2) Dual purpose: such as grain and also general cargo.
3) Special purpose: such as tankers, refrigerated ships.

The development of trade in specific types of cargo has created special types of ships for the carriage of such cargoes in bulk. The bulk grain carrier differs little in external appearance from the general cargo ship but it is, however, necessary in ships intended for this service to make special provision in the cargo spaces to ensure that the grain cargo does not shift. These arrangements generally consists of longitudinal partitions in the cargo spaces.

The world demand for oil has led to a special type of ship, the oil tanker. There has been a great increase in the size of these ships and there are now tankers capable of carrying in excess of 275,000 tons deadweight. In this one deck type of ship the machinery is placed aft and the structural arrangement in way of the oil cargo space is to divide the ship in the athwartship direction into three by means of two longitudinal bulkheads. The cargo space is also divided into tanks in the longitudinal direction by means of transverse bulkheads. Pump rooms are provided with powerful machinery to deal with cargo; the tanks have small oil-tight hatches on the deck and adequate ventilation of the cargo tanks is provided to take away gas which may be formed.

The large bulk carrier was initially developed in the Great Lakes where single deck ships were used for the carriage of ore. This is a trade which has expanded considerably and the carriage of ore on the high seas today employs many straight ore carriers and ore/oil carriers. Many features common to ore-carriers such as large hatchways, clear holds and steel hatch covers have been adopted in dry cargo ships.

Food ships constitute an important section in the passenger-cargo type. These vessels have extensive insulated spaces for the carriage of meat, fruit and dairy produce and are provided with refrigerating machinery to maintain the spaces at the temperature required for the cargo being carried. Volume for the cargo is an important factor since the cargo carried has a fairly high stowage rate. The speed of these refrigerated ships is higher than the ordinary cargo ship as it is desirable to keep perishable cargoes at sea for as short a time as possible. Because of their speed these vessels attract passengers who are well catered for by quite luxurious accommodation.

A not unimportant type of cargo ship is one that carries a deck timber cargo. The timber requires to be properly secured in order to prevent it shifting when the ship is at sea. As the cargo is stowed at a high position special attention must be paid to the question of stability. There are special regulations in the Load Line Rules for such ships.

Container Ships

The reduction of cargo handling costs has become increasingly important and one way of achieving this is by new cargo handling concepts such as unit transport techniques. The application of unit transport is being continually extended and container services on short and long distance routes have been operated.

Containers have been used for specialized cargoes such as furniture for many years and in recent years the idea has been applied to a greater range of commodities and to larger units. Container services originated in America during the second world war and the U.S.A. has considerable experience in their operation. This type of service has been very successful round the Australian coast.

Specialized services of this kind require specialized ships and the first of these to be built in the United Kingdom was by John Redhead and Sons at South Shields. Each ship carries 148 standard containers having dimensions 30 ft by 8 ft by 8 ft. Of these 110 are stacked, mostly three high, in the cellular holds and the remaining 38 on the hatch covers. The whole of the cargo space in such a container ship is taken up by sectionalized holds containing cellular compartments. Instead of cargo of assorted shapes, weights and sizes, the payload consists of containers of standard dimensions which slot into individual cells. By bulking the cargo into unit loads in this way, the loading and discharging operations can be fully mechanized.

The container ship is the waterborne link in the system of transportation wherein cargo is transported within uniform containers interchangeable by different forms of transportation.

Details of the foregoing types and of the vessels shown in the family tree of Table 3.4. including those under the heading *Subsidiary* are given later.

Although much prominence is given to large ships it should not be forgotten that the small ship plays an important part in the world of shipping. Without lightships the sea lanes could not be kept safe; without dredgers there would be no deep and wide navigable lanes to harbour entrances; without tugs it would in many cases be impossible to dock the ship.

Requirements

Many of the requirements which a ship must fulfil are common to all types, passenger ships, cargo ships and ships designed for a special purpose. It is desirable that some of these requirements be enumerated.

The first requirements is that the ship when at the maximum draught permitted can carry the load for which it was designed. The mass of the ship itself, called the light mass, is made up of several items including the structure, propelling machinery and all fittings to make the ship complete and ready for sea. The load is called the deadweight or dead load and includes cargo, fuel, fresh and feed water, stores, passengers and baggage, crew and effects. The simple relationship arising from this is:

$$Displacement = Light\ Mass + Deadweight\ or\ Dead\ Load$$

The second requirement is that the ship shall be stable in all normal conditions of loading. By stability is meant that when the ship is inclined from the vertical by some external force it will return to the vertical when the external force is removed.

The third requirement is that the hull will have sufficient strength by the disposition of the material of the structure to withstand all the stresses that may be encountered in service and remain in a condition of seaworthiness.

The fourth requirement is that the desired speed be obtained with the minimum use of power and consequently low consumption of fuel. This means designing efficient propelling machinery, and propeller and the design of an underwater form to give the lowest possible resistance to motion through the water.

In addition to the general requirements which apply to all ships there are many special requirements which depend upon the type of ship under consideration. All are necessary to produce an efficient ship. The main technical factors that influence efficiency are size, speed, type of machinery, method and materials of construction and loading and discharging facilities; to these must be added the accommodation and amenities provided for the crew.

However, overriding all these are considerations of safety. The fundamental objective in ship design is to ensure that the ship is seaworthy. The term "seaworthiness" covers a wide range of supplementary qualities. The ship must be capable of floating in the worst seas which will be encountered; must be expected to remain afloat following damage, as even with the highest standard of seamanship and with the assistance of the most modern devices, accidents at sea may still be expected to take place. It is worthy of note that not all types of ship are expected to have the same margin of safety against damage at sea; loss of goods is at any time regrettable, but the loss of lives is, of course, even more so, and thus passenger ships are required to withstand with safety a greater extent of damage to the hull than a cargo ship of the same size.

Adequate stability is an essential attribute of the seaworthy ship. More than external forces are considered in this connexion; despite all precautions cargoes may shift at sea in heavy weather and the ship must have a reasonable degree of reserve stability. For the stability of a ship at large angles of inclination an important factor is freeboard —the amount of the ship above the water. It is the freeboard which governs the angle at which the deck edge comes awash as the ship is inclined from the vertical. Freeboard affects extensively the range of stability, that is the angle of inclination at which the ship would become unstable and capsize. A seaworthy ship must have all openings in the sides and weather deck secured against ingress of water; the propelling machinery and steering gear must be efficient and reliable.

It should not be forgotten that for all the advances in recent decades the condition of safety and success at sea is still unceasing vigilance by all concerned. Flesh and blood have contributed more than steel and oil to maritime achievement.

Crew Accommodation

Crew accommodation in merchant ships today is an important factor in the general design of ships. The provision of proper accommodation for the crew in British ships is a matter with which the Department of Trade and Industry is concerned and certain minimum requirements are stipulated in the regulations. Many owners take a great deal of interest in crew accommodation and in a large number of ships the standard is well above that required by law. Considerable emphasis has been placed on the arrangement, furnishing and decoration of the crew accommodation and on the amenities and recreational facilities provided for off-duty periods.

The regulations relating to accommodation require the provision of suitable spaces above the load waterline either amidships or aft. The spaces must be readily accessible and afford protection from the weather and the sea. The accommodation must have an approved system of lighting, both natural and artificial and be adequately heated and ventilated. Separate sleeping quarters must be provided for officers, petty officers, apprentices and ratings of the deck, engine room and catering departments. The number of persons accommodated in sleeping rooms for officers and petty officers is one person per room; for apprentices not more than three persons per room; for other ratings in cargo ships not more than four persons per room. Each category must also be provided with mess rooms, wash rooms and sanitary arrangements for its exclusive use.

It is now quite common to find a high proportion, if not all, of the ship's complement housed in single-berth rooms and with well appointed recreation rooms and smoking rooms. It can indeed be said that those who go to sea today are excellently housed.

Propelling Machinery

The selection of the type of propelling machinery to be adopted for merchant ships is based upon its reliability in service and the cost of operation and maintenance.

The post war developments in the direct drive diesel engine virtually brought to an end the era of the reciprocating steam engine. Licences to manufacture diesel engines were secured by many shipbuilders and their associated marine engine works. The economy of the diesel was a great asset and it was not long before the diesel engine had taken over where the steam reciprocating engine had once been supreme.

For the average general cargo ship, steam turbines due to their relatively high fuel consumption are not economically attractive. Consequently in the low and medium power ranges the diesel engine with its low fuel consumption has taken over. Most motor cargo ships have machinery of the slow running type directly connected to the propeller shaft. Today's trend is to have more power in less space and current practice is to turbocharge existing standard type diesel engines.

Steam turbine machinery even in its most advanced stage is unable to compete with the low fuel consumption rates of the diesel. Generally, however, the total weight of the installation is less than that of the corresponding motor ship and maintenance and repair bills are lower. To make the steam turbine more competitive efforts are being made towards reductions in initial cost and fuel consumption. Steam turbines are continually being improved and a re-heat turbine has been installed in one of the world's very large tankers. The steam turbine is likely to continue to be used in ships where reliability and freedom from vibration are important factors—passenger ships.

There is ample evidence to show that the marine engineering industry moves with care and caution when new developments are under consideration. Events which occur before the correct time such as the gas turbine propulsion development, are generally regarded with a modicum of suspicion. With the large, slow-running, direct coupled marine diesel engine, turbocharging and design improvements giving reduced overall weight per horse-power have been successfully applied in gradual stages. The result of this is that such engines are now available for powers in excess of 50,000 horse-power.

For the average general cargo ship today the frequently accepted proposition seems to be a turbocharged, two stroke direct drive diesel using exhaust gases for a waste-heat boiler and burning residual fuels.

The medium speed, geared diesel has become established in

recent years. The power-to-weight ratio of a slow speed direct drive diesel is high and the introduction of a diesel that operates at medium speed and with reduction gear to provide a lower propeller speed resulting in a power weight ratio of about one-fifth of the slow speed diesel has had much to do with the increasing popularity of medium speed diesel engines. Indeed an increasing proportion of the marine diesel market is claimed by the medium speed diesel. This popularity is being enhanced by the production of engines having a much greater power per cylinder than hitherto.

Gas Turbines

An extensive research programme by the US Maritime Administration on marine propulsion systems and the adaptation of industrial gas turbines for marine use has served to focus attention on the vast untapped potential of the marine gas turbine engine.

In 1956, a gas turbine was installed in the liberty ship *John Sergeant*. This was the first gas turbine-powered dry cargo ship designed to use chemically-treated residual fuel oil. Although this ship burned heavy fuel for several thousand hours over a period of three years the plant's total economics were not competitive. However, manufacturers have been developing turbines with higher compression ratios and higher firing temperatures so that the efficiency of the plant could become competitive with steam and diesel power. The availability of different engine outputs indicates that the industrial gas turbine has the potential to achieve a large measure of acceptance in merchant shipping. There is also the aircraft-type marine gas turbine, the outstanding characteristic of which is its superior power to weight ratio.

Ship propulsion is undergoing a significant transformation. A decade ago, the steam turbine and slow speed diesel were the basic commercial choices for ocean-going ships. Today the shipowner has available to him many competitive power plant alternatives in the form of conventional and reheat cycle steam turbines; slow and medium speed diesel engines, and industrial and aircraft gas turbines. This profusion of power plants creates a challenge in evaluating the many factors which have to be considered when selecting a power plant. Some of these are the power output required, the space required for the propelling unit, the relative cost and adaptability to central and remote control, maintenance costs, fuel costs and the important factor of reliability.

In the years to come, the selection of propelling units is certain to become more complex, while cost will remain the primary concern.

Marine Nuclear Power

Commercially competitive nuclear power plants for merchant ships have yet to be demonstrated. There are at present nuclear powered ships at sea but none make any pretence to economic operation.

There are many aspects of the subject of nuclear powered ships and consideration has to be given by the ship designer, shipowner, shipper, port authorities, international requirements, classification societies and marine underwriters. A very brief summary of each of these is as follows:

Ship Designer: Although a marine nuclear installation may be heavier than conventional propelling machinery without fuel, the nuclear powered ship of comparable size will be able to carry more cargo than the conventional ship on long voyages at relatively high speed. Realization depends upon the development of an economic marine unit.

Shipowner: Any propulsion machinery for merchant ships is judged by the cardinal requirements of reliability in service and cost of operation and upkeep. There is no economic sense in adopting a new form of marine propulsion unless it is profitable.

Shipper: The over-riding interests of the shipper are that freights should be low and voyages safe and swift. The shipper is not directly concerned with how the shipbuilder and marine engine-builder achieve the end product but is very much concerned with the results produced.

Port Authorities: The requirements for cargo handling and the precautions necessary while a ship is in port will differ for nuclear ships from that for the traditional type of ship. However, it appears from investigations already made that provided suitable precautions are taken nuclear vessels could be dealt with at many ports.

International Requirements: A number of recommendations concerning safety aspects of nuclear ships are included in the current Safety of Life at Sea Convention. Present day developments indicate that a number of points will arise for consideration including the manning of such ships.

Classification Societies: Consideration has been given by the classification societies to hull materials, hull strength and collision and grounding protection for both hull and machinery of nuclear ships. Lloyd's Register of Shipping has provisional rules for the classification of nuclear ships.

Marine Underwriters: It is understood that marine underwriters do not anticipate serious problems to arise when a nuclear ship is proposed for insurance. Insurance can be made available during construction and also while the ship is being operated.

CHAPTER 4

Merchant Ship Types

The United Kingdom still has one of the largest shipping fleets in the world although its shipbuilding industry has been, for some time, steadily decreasing in strength. Britain has always been a trading nation and the development of shipping was essential to becoming the premier industrial nation in the 19th Century. Since that time the mercantile lead in the world has not been paralleled with technological supremacy.

The first practical steamship was the *Charlotte Dundas* built by Symington and tested in the Forth–Clyde Canal where in 1803 it travelled for about 20 miles at six miles an hour. The first steam-powered British ship to proceed to sea was Henry Bell's *Comet* which plied as a passenger vessel in the Clyde estuary. These early steamships were all paddle steamers until the introduction in 1838 of the marine screw as a means of propulsion. The eventual supremacy of steam over sail was brought about by the compound engine, the screw propeller and the iron hull.

British shipping is facing the challenge of world competition by a radical reorganization. A modern fleet is imperative in view of the change from the conventional cargo liners and tramp ships to bulk carriers and container ships. There is a distinct tendency towards new specialist services in shipping which will make it possible to use large ships efficiently.

As to the future there are three main considerations:
1) *The Nuclear Ship.* The United Kingdom has postponed procedure in this field until a reactor is available that offers economy in operation which compares with that of the present ship using oil.
2) *Automatic Control Systems.* These would enable all control to be made from the navigating bridge. The Japanese introduced this kind of automation in 1961.
3) *Automatic Steering.* This would eliminate constant attention of crew.

Passenger Liners

The requirements of the various services on the regular passenger routes differ appreciably. For example the Atlantic has produced a unique class of ship, the mammoth Queens and the like. Such ships, above all others, are subject to the ever-increasing pressure of air transport.

In ships which are predominantly passenger-carrying, the space available for cargo is of minor importance. The controlling factor is space for the number of passengers it is intended to carry. This involves the provision of cabins and public rooms such as dining rooms, lounges as well as cinema, theatre, swimming pool, shops, etc. and the greatest possible extent of deck promenade space. Many passenger ships engage in cruising in the off-season and in some instances this may entail a portion of the accommodation being interchangeable.

A feature of passenger liners is the extent of superstructure above the uppermost continuous deck. The addition of several tiers of superstructures creates problems of strength and stability but the adoption of aluminium alloy for the superstructures has gone a long way to resolve such problems.

The comfort of passengers is most important and has received a great deal of attention including air conditioning and the fitting of anti-rolling fins. The highest standard of comfort together with impeccable service was provided by the then Cunard liners *Queen Mary* and *Queen Elizabeth* on the Atlantic service. They were withdrawn from service for the simple reason that they did not pay their way. The days when crossing the Atlantic by a huge liner such as the *Queen Mary* made any kind of economic sense are gone. The voyage is too short and too uninteresting for a holiday; too long for those travelling on business; too expensive for anything at all.

The *Queen Elizabeth 2* launched on the Clyde in September 1967 represents the latest concept of the giant passenger ship and may indeed be the last of the mammoth passenger ships. In recent decades many social and technological changes have taken place and these have brought about a different conception of the function of a passenger liner. In 1938 the *Queen Mary* was the fastest means of transatlantic transportation but today the jumbo jet has changed all that. The *QE2* was intended for service on the North Atlantic run during the summer and as a one-class ship for world cruising. The advantage this ship has is in making available what aircraft cannot provide—a luxury hotel way of travel.

Some particulars of the *Queen Elizabeth 2* are given in Table 4.1.

Passenger-Cargo Ships

The number of large passenger liners is being reduced due mainly to air competition and many of these ships are now engaged in cruising. It is claimed that a cruise is the purest form of civilized idleness and that all the fun of a cruise can be obtained aboard a liner on regular passenger runs.

The passenger cargo liners are in a rather different category. They carry considerable quantities of cargo and fewer passengers than the pure ocean liner; the exact quantity of each being variable. The accommodation in these vessels is not generally as luxurious as that of the fast ocean liner but is nonetheless of a high standard.

Table 4.1.—Queen Elizabeth 2

Length overall	294·0 m
Breadth	32·0 m
Overall height of ship to top of funnel	61·4 m
Draught	9·9 m
Service speed	28·5 k
Passenger capacity	2,025
Propelling machinery	Double reduction geared turbines
Power output	82,000 kW
No. of propellers	2 (six-bladed)
No. of boilers	3
No. of decks	13
No. of funnels	1
Swimming pools	2 Outdoor 2 Indoor
Lifts	22
Classes	One class

Hull Subdivision

The standard of safety applicable to passenger ships has been determined by various International Conferences. Such a Conference in 1960 on Safety of Life at Sea embodied the latest recommendations and requirements in a new Convention. One of the means of protection from the hazards of the sea is to have a system of watertight compartments such that should the hull be pierced there will be a good chance that the ship will remain afloat for some time. The benefits of subdividing a ship by transverse bulkheads have been appreciated since very early times. The current requirements are embodied in the 1960 Convention mentioned above.

The effects of damage which destroys the watertightness of a ship's hull are often progressive. They depend upon the extent of the damage in relation to the ship's internal arrangement and the type and extent of any cargo or other material in the damaged compartments. The extent of flooding in any compartment depends, of course, upon the amount of empty space in the compartment.

For passenger ships—defined in the Merchant Shipping Acts as those carrying more than 12 passengers—certain standards of subdivision are stipulated. The degree of subdivision into watertight compartments varies according to the nature of the service of the ship and its length on the load waterline. The highest standard of subdivision is applied to a ship primarily engaged in the carrying of passengers and is gradually reduced in order to suit a ship which comes close to the purely cargo type.

The method adopted is to determine a line beyond which the ship should not sink and then ascertain the position and length of the compartment which when flooded will cause sinkage to that line. Such a line is known as the Margin Line and is a line drawn parallel to and 76 mm below the surface of the bulkhead deck at the ship's side. The bulkhead deck is the uppermost continuous deck to which all transverse watertight bulkheads are carried. The effective volume of a compartment for flooding depends upon the permeability of the compartment. The percentage volume of a space that can be flooded is known as the permeability. If a compartment is filled with sponges the permeability would be very high, probably 90 per cent. On the other hand if the compartment is filled with baulks of timber, square in cross-section and closely packed, the permeability would be very low, probably 20 per cent.

The international regulations for damage stability are exacting. They require that if two adjacent compartments are flooded the ship will not only stay afloat but the angle of heel is limited to a maximum of 7 degrees.

Hull Construction

In general a double bottom is required by the regulations to extend from the fore peak bulkhead to the after peak bulkhead and the inner bottom plating continued to the ship's sides so as to protect the bottom to the turn of the bilge. This protection is deemed satisfactory if the line of intersection of the outer edge of the margin plate with the bilge plating is not lower than a horizontal plane passing through the point of intersection with the frame line amidships of a transverse diagonal line inclined at 25 degrees to the base line and cutting it at a point one-half the ship's moulded breadth from the middle line as shown in Figure 4–1(a). The inner bottom plating is frequently carried out to the ship's side as in Figure 4–1(b). The double bottom in addition to providing a measure of protection in the event of damage to the outer shell by grounding has the great advantage of making available spaces for the carriage of oil fuel, fresh water and water ballast.

Protection of Ship against Fire

The three basic principles underlying the regulations are:
1) separation of the accommodation spaces from the remainder of the ship by thermal and structural boundaries.

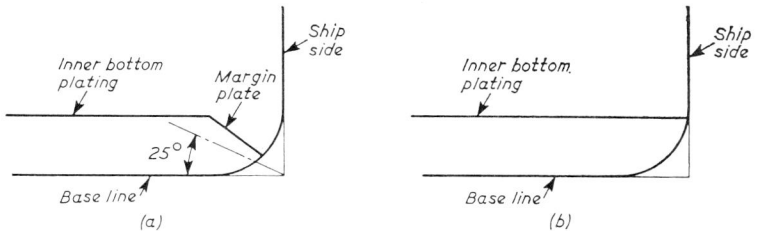

Fig. 4–1 (a) and (b)—Double bottom.

2) containment, extinction, or detection of any fire in the space of origin.
3) protection of means of escape.

The hull, superstructure and deckhouses are subdivided into main vertical zones by fire-resisting bulkheads not more than 40 m apart.

The accommodation spaces and service spaces have to be constructed in accordance with one of the following methods of fire protection:

Method I. The construction of internal divisional bulkheading to be of fire-retarding divisions.

Method II. The fitting of an automatic sprinkler and fire alarm system for the detection and extinction of fire in all spaces.

Method III. The subdivision of the spaces by fire resisting divisions, together with an automatic fire detection system and a restriction of the use of combustible materials and furnishings in these spaces.

Special arrangements are required in the machinery spaces.

Cross Channel Ships and the Car Ferry

In general these vessels form the link between railheads, the sea routes being planned so that connecting rail services are available from the terminals at one or both ends of the voyage.

Cross channel ships usually have a speed in excess of 20 knots and consequently are high powered. For good manoeuvrability they have twin screws and frequently a bow rudder and in a number of cases a propulsion unit in an athwartship tunnel forward in order to provide sideways thrust for rapid turning. Stabilizers are also generally fitted.

During the past few years new routes have been opened and new ships engaged in the conveyance of motor cars, coaches and commercial transport and large numbers of passengers. The car ferry type of vessel is now an important vehicle in sea transportation.

The cross channel ship for day service has large public rooms in lounges, restaurant and/or cafeteria, bars etc. arranged generally on three or four enclosed decks. Extensive promenade space is also made available.

In the passenger-vehicular ferry the arrangement below deck is subordinated to garage space. On the upper decks passenger accommodation is provided. These vessels can operate as drive-on drive-off types of ship and may have hydraulically operated doors at the bow and stern.

The modern ferry is a ship to cater for three dimensions—passengers, motor vehicles and cargo packed in containers. All three are carried in isolation from one another on different decks. Comfort, good food, drinks and duty free shops on board are the keynote of modern ferries for passengers and rapid door-to-door delivery, the aim of the cargo exporters. In general the ferry today is one class as it simplifies design and keeps down costs. In car ferries it is customary to have a large main restaurant as well as a snack bar to provide tea, coffee and light refreshments.

The profile of a car ferry is shown in Figure 4-2.

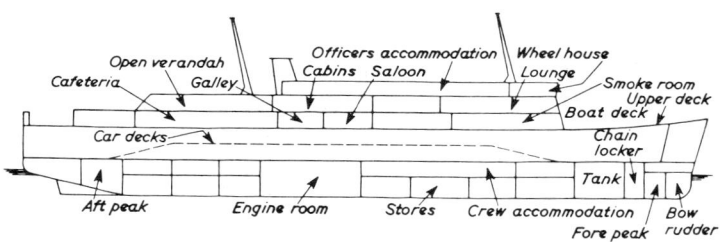

Fig. 4-2—*Profile of a car ferry.*

Cargo Carriers

There are many different types of cargo ships and the type to which any particular ship conforms is dictated by the duties which the vessel has to perform. In these the passenger element is not altogether eliminated since many cargo carriers have cabin accommodation for passengers up to a number not exceeding 12. Beyond this number the ship would require a passenger certificate.

Cargo carriers may be put into three main divisions:

1) General trading
2) Dual purpose—such as grain and also general cargo
3) Special purpose—such as tankers, refrigerated ships, container ships.

A brief description of the salient features of the most important cargo ship types is given below.

Oil Tankers

The growth in the size of oil tankers has been outstanding. At the end of World War II the average deadweight of tankers built throughout the world was 14,000 tonnes. In a decade later the average size had doubled. At the present time designs are being considered for the tanker of 1,000,000 tonnes deadweight.

The oil tanker is one of the few ship types in which the cargo directly rests on the skin of the ship, without the interposition between it and the sea of any other structure such as a double bottom.

The reasons for the rapid increase in tanker size are both political and economic. From the viewpoint of the latter, related to the quantity of oil carried, one large tanker can probably replace three smaller ones and consequently make considerable savings in crew costs. Again one large power unit is more economical in operating costs than three smaller units. Many other costs including dues and pilotage do not increase in direct proportion to the deadweight and thus there is a saving with the large vessel. However, very large ships necessitate correspondingly large dry docks, harbours and repair facilities to meet their requirements.

The carriage of petroleum in bulk necessitates special precautions being taken to meet the difficulties associated with the transport of such a cargo. The propelling machinery is fitted aft and in way of the machinery space there is a double bottom for the oil fuel etc. The modern tanker has the cargo space subdivided longitudinally by twin oiltight bulkheads and transversely by a number of oiltight bulkheads throughout the length. This gives several sets of three tanks. To minimize the risk of leakage of oils or vapour into other compartments, a pair of bulkheads, forming cofferdams, are fitted at each end of the oil cargo range. These longitudinal and transverse bulkheads contribute to the longitudinal strength and general stiffness of the hull.

The cargo oil-pumping arrangements are quite extensive since a number of grades of oil may have to be loaded, transferred from tank to tank and discharged by a pipe network without risk of contamination of one grade by another.

Tanks for heavy oils, molasses or other viscous fluids are fitted

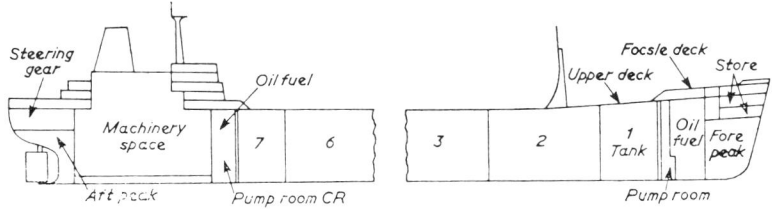

FIG. 4-3—*Profile of an oil tanker.*

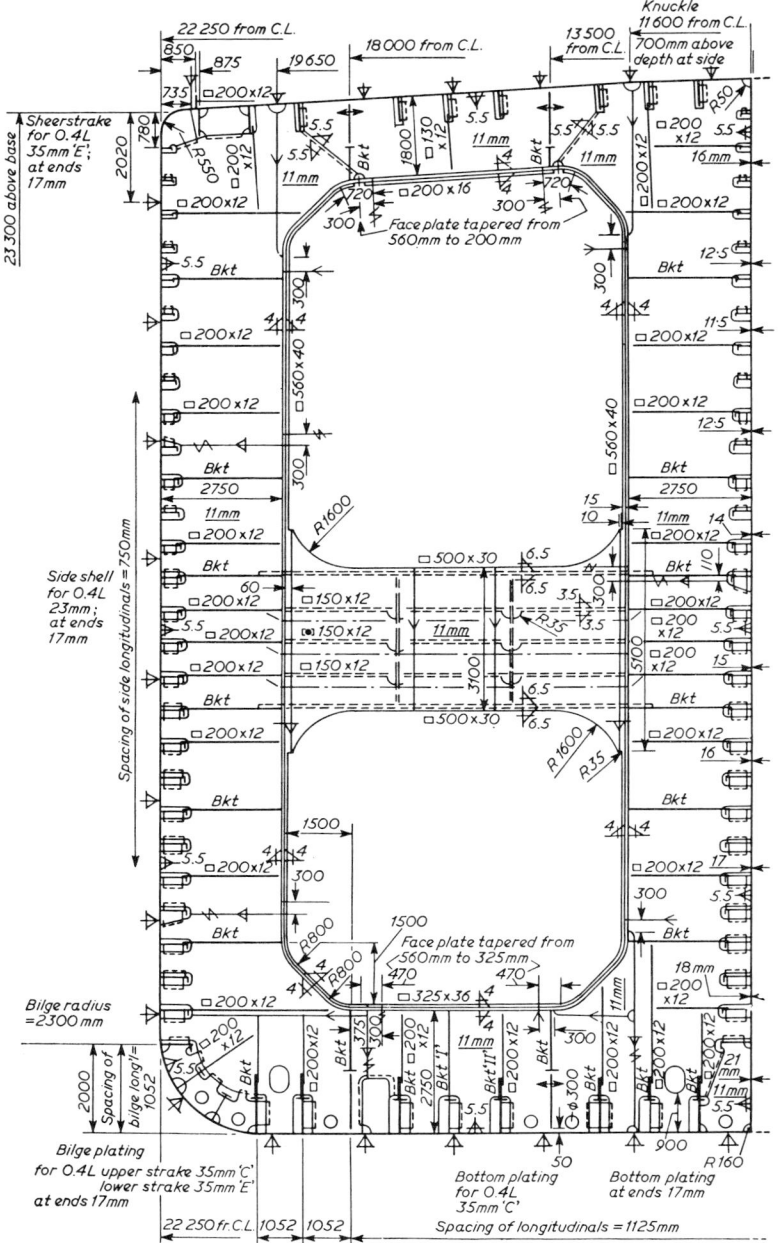

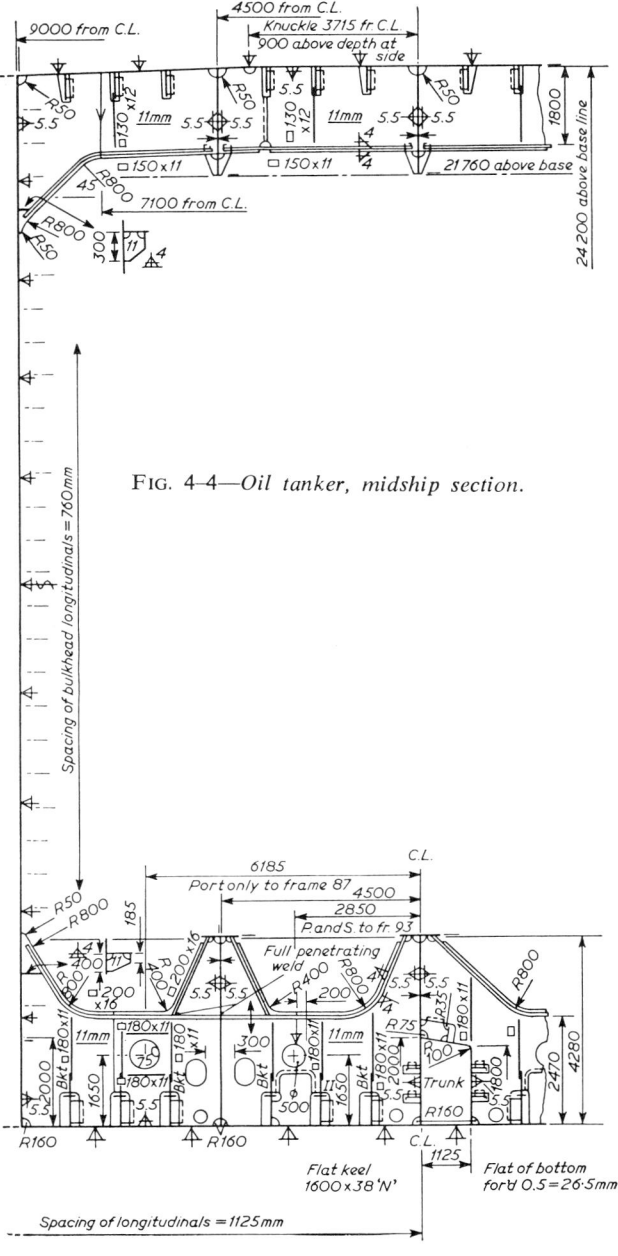

Fig. 4-4—*Oil tanker, midship section.*

with heating coils so that the cargo may be adequately liquefied to be able to run freely to the pump suctions.

As an explosive mixture is formed when oil vapour comes into contact with air it is essential to ensure that foul gas which may have accumulated from the oil has a ready means of escape.

In order to reduce the still-water bending moment and thus make lighter scantlings possible the large tanker today has certain tanks near amidships arranged as permanent empty spaces.

With regard to the provision of a desirable margin of stability, under sea-going conditions, the oil tanker unlike some other types of cargo ships is under the control of the designer. With the normal dry cargo ship the loading and the resulting position of the vertical centre of gravity is not regulated by the ship designer. In the tanker, however, the mere filling of tanks makes it simple to determine the position of the centre of gravity above the keel and thus simplify the stability investigation. Nevertheless, in the process of loading or discharging liquid cargoes there can be a reduction in the metacentric height (GM) which is the vertical separation of the centre of gravity (G) and the transverse metacentre (M). With reference to a ship's initial stability or stability in the upright condition:
a) If G is below M the ship is in stable equilibrium
b) If G is above M the ship is in unstable equilibrium

It is thus easy to see how important the relative positions of the centre of gravity and the transverse metacentre are as affecting a ship's initial stability.

The problem of stability is considered in some detail in Chapter 11.

The profile of a tanker is given in Figure 4–3 and the midship section is shown in Figure 4–4.

Bulk Carriers

The large bulk carrier was developed initially in the Great Lakes where, by the beginning of the twentieth century, large single deckers were used for the carriage of ore. Bulk carriers are in general single deck ships with machinery aft. There is a separate section in Lloyd's rules which deals exclusively with this type of ship. These vessels are designed basically for the transportation of grain, coal or ore cargoes. The solid bulk cargo is such that it can be loaded into the ship's hold by gravity and discharged by grabs, a conveyor system or as in the particular case of grain, by suction.

The bulk carrier design as now evolved can be used for the transportation of a wide range of cargoes from motor-cars at one end of the scale to oil and ore at the other. Competition in the raw commodity markets of the world has brought full realization of the economics of shipping in bulk. The growth in the size of the fleet of this distinct type of ship in recent years has been considerable as also in the size of such ships. In the decade from 1960 the average bulk carrier has increased in deadweight or dead load from 20,000 tonnes to in some cases in the region of 100,000 tonnes. The growth of the bulk carrier is similar in many ways to that of the tanker but not so rapid. It has, however, led to the combination ship in the oil/ore carrier and the oil-bulk-ore (OBO) carrier.

An outstanding example of this type of ship is the *Furness Bridge* with a deadweight or deadload of 173,200 tonnes. This vessel when delivered in 1971 was the largest OBO carrier in the world.

A detailed description of this ship is given in "Shipping World and Shipbuilder" for September 1971, Vol. 164 No. 3861.

Over the years several structural arrangements have been adopted in this type of ship and of these the most popular is that with the double bottom sloped up to form a wing tank, and a topside tank extending from the hatch side to the side of the ship as shown in Figure 4–5.

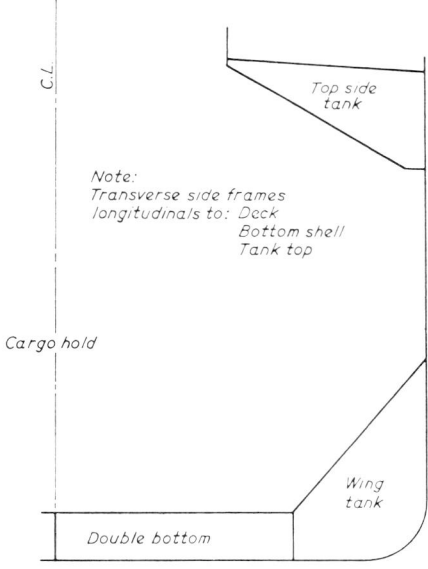

FIG. 4–5—*Bulk carrier, midship section.*

In ships designed for the carriage of heavy cargoes such as iron ore special provision is made. These vessels generally have two longitudinal bulkheads as on a tanker. The ore is carried in the centre compartment between these bulkheads and on top of a deep double bottom as is shown in Figure 4–6. The depth of the double bottom is such as to keep the centroid of the ore high enough to prevent the ship being too stiff in a seaway. The stowage rate of iron ore is about 0·48 m³ per tonne.

In the ore and oil carrier the main structure of the ship has to be in accordance with the ore standard and in addition the bulkheads and other appropriate parts of the structure have to be equivalent to oil tanker standard. The transverse section of this type of ship is

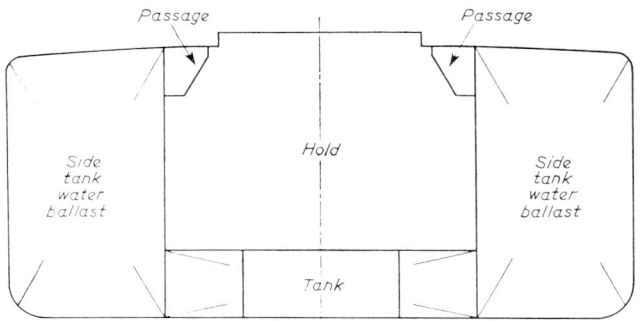

FIG. 4–6—*Ore carrier, transverse section.*

given in Figure 4–7 from which it will be seen that underdeck passages for the safe movement of the crew about the ship are provided.

Unlike the oil tanker a bulk carrier must enter a port to be loaded or unloaded. It cannot unload at the end of a jetty extending into deep water with the ease of the oil tanker. Its size is subject to the dimensional limitations of docks and harbours.

Bulk Carrier Types

The principal differences between bulk carrier types are as follows:

Tanker	No hatch covers.
Bulk Carrier	Large hold volume for low density cargoes.
Ore Carrier	Small central hold with deep double bottom. High density cargo of 0·35 to 0·50 m³/tonne.
Ore/Oil Carrier	This is virtually a tanker with a hatch in way of the central tank through which high density ore can be loaded. Ore and oil are not carried simultaneously due to explosion risk.
Oil/Bulk Ore Carrier	This is a bulk carrier in which the structure is reinforced to deal with oil and high density cargo. Frequently alternate holds such as Nos. 1, 3, 5 and 7 are smaller and these only would be loaded for ore transport.

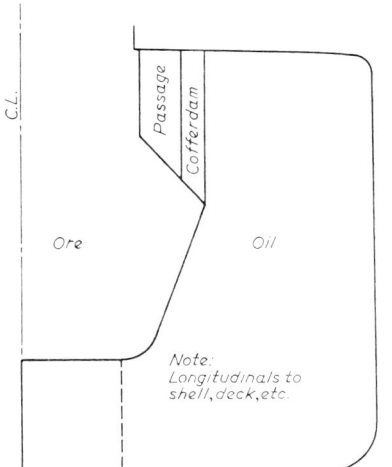

FIG. 4-7—*Ore/oil carrier, transverse section.*

Refrigerated Cargo Ships

Food ships constitute an important section in the passenger-cargo type. These vessels have extensive insulated spaces for the carriage of meat, fruit and dairy produce and are provided with refrigerating machinery to maintain the spaces at the temperature required for the cargo to be carried. Refrigerated cargo, because of its perishable nature, and because of the cost of running refrigerating machinery for protracted periods requires fast ships. Due to their speed these vessels attract passengers who are well catered for by quite luxurious accommodation.

Volume for the cargo is an important factor since the cargo carried has a fairly high stowage rate. Chilled beef stows at about 3·5 m^3 per tonne; frozen beef stows at about 2·6 m^3 per tonne; bananas at about 4·2 m^3 per tonne. Due to the depth of insulation the volume of the cargo spaces is quite substantially reduced. In the case of chilled meat the reduction from the moulded volume could be 35 per cent and the corresponding figure for frozen meat is about 25 per cent.

Some classes of cargo such as mutton and pork can be frozen hard without damage to their food value, others such as beef cargoes generally arrive in better condition if only chilled. Chilled beef is hung from hooks and chains and there is about one-third of a metre clear below the meat to permit circulation of air. The height of storage of frozen meat should not exceed 6 metres to avoid crushing lower tiers.

Bananas are stowed in bins, generally two tiers placed vertically and one tier horizontally. Each cargo space is divided into bins; permanent uprights are fitted at about 3 m intervals fore and aft and athwartships. The sides of these are slotted to take portable battens.

Container Ships

The container principle is neat and simple. A number of small packages are stowed inside one large container which is then carried around the transport system as one unit. This saves a great deal of time and man-handling in moving goods by rail, road and sea. It permits much more mechanization of the handling processes since all the items are in standard sizes. Container ships can load and unload as much as ten times more quickly than the conventional cargo ship.

The container business is developing at a fast pace on the main trade routes of the world. In the roll-on, roll-off form it is an accepted part of the shipping scene and companies have been formed who have introduced cellular ships specially designed for the service. Although small numbers of containers have been carried in conventional cargo ships for many years it was not until 1956 that the first container ships were constructed.

The container ship is equipped to carry containers in the holds and on the weather deck by means of special structural arrangements and devices. Within the holds of such vessels there is a cellular structure of angle bars forming container guides into which the containers are stowed one on top of another. The containers carried on deck are carefully lashed down to ensure that they will not shift. These vessels are kown as cellular container ships. The only movement of the container within the ship is vertical and thus loading and discharging is carried out by vertical movement without horizontal movement. The cells are generally arranged fore and aft and in groups the number depending on the breadth of the ship. The transverse width of the groups may be 80 per cent of the breadth of the ship. Consequently large hatches are essential to use the holds to the best advantage. The fore and aft groups between bulkheads are separated by heavy web frames or similar structure to act as support to the cells and also give rigidity to the vessel. Each cell circumscribes one container stack and extends from the hatch coaming level to the tank top.

Figure 4–8 shows the midship section of a cellular container ship.

It has been said that the "wheel" was the basis of great advances in the entire field of communications. Today the "box" has arrived and with it a great new concept in the national and international carriage of goods. The container is a box, with doors at one end or at the side, into which the merchandise is packed. The materials at present used in the construction of containers are steel, stainless steel, aluminium, plywood, plastics including glass reinforced plastic (GRP). The containers are, in general, built to the International Standards Organization (ISO) requirements. The standard nominal sizes of series I freight containers are 6,000, 9,000 or 12,000 mm in length and all having a cross section of 2,435 by 2,435 mm.

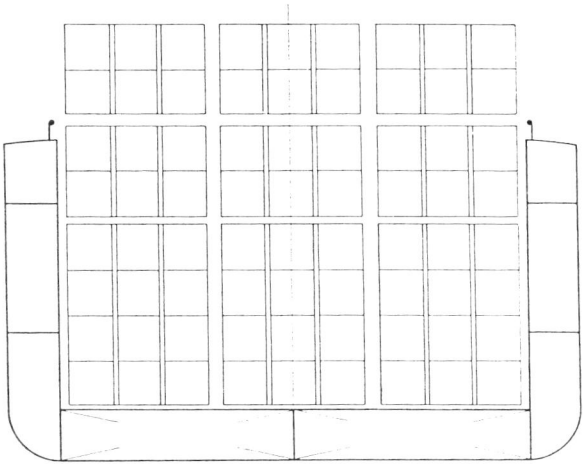

FIG. 4–8—*Container ship, midship section.*

Tugs

In general, ships are designed for a specific duty and seldom are called upon to perform any other. The tug, however, is expected to render services which are many and varied, and the service in restricted waters demands high standard manoeuvring qualities. Whilst all other ships are powered solely for themselves, the efficiency of the tug as such is dependent on the amount of power the vessel can transmit through a tow rope to some other vessel.

Generally tugs may be classified according to their duties as follows:
1) River   2) Harbour   3) Coastal   4) Ocean going and Salvage.

The basic requirements for all tugs are stability under all conditions of operation, manoeuvrability and adequate towing power. As to manoeuvring the Kort rudder and Voith Schneider propeller have made a very high standard possible. The controllable pitch propeller and bridge control of the propelling machinery are now fairly standard.

Even in sheltered waters the operation of tugs is not without its difficulties. The major risk for the ship-handling tug is that of being dragged sideways by the towline until the vessel heels over and possibly capsizes; this can be overcome by a towing hook that releases under a predetermined load.

In the ocean-going and salvage tugs the main features are power, sea-going qualities and radius of action. The conditions for towing in ocean service are very different from that encountered in the other tug range. In bad weather a long tow-rope is essential and to prevent breakage of the tow-rope a towing winch is used. Such a winch takes a pre-determined load and if this is suddenly exceeded the winch automatically pays out wire; when the excessive load is relaxed the winch takes in the wire. Ocean-going tugs carry extensive salvage and fire-fighting equipment. A comprehensive range of navigational equipment is provided together with extensive floodlights and a searchlight.

The midship section of a tug is shown in Figure 4–9.

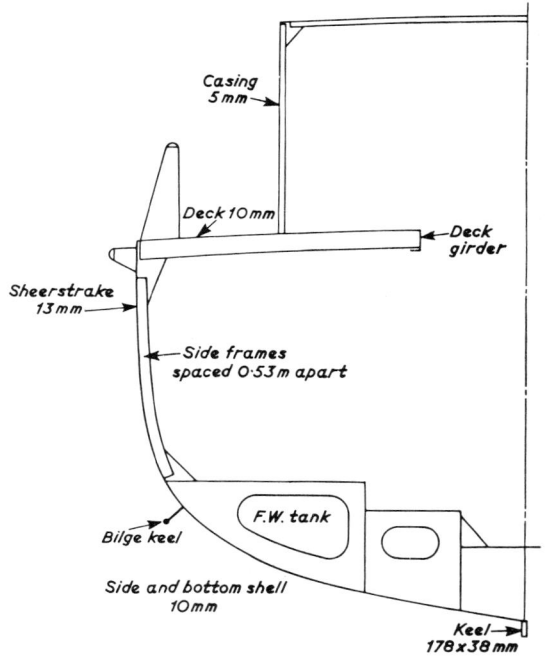

FIG. 4–9—*Midship section of tug.*

Liquid Gas Carriers

For ships that carry liquefied gases the range of cargoes covered is quite formidable and involves ships for both pressure and refrigerated transportation. The greater part of liquefied gas traffic at present is in liquefied petroleum gas (LPG) and the rate of usage in many countries is increasing rapidly.

Butane is the simplest of the common petroleum gases to be transported in liquefied form as only low pressures are necessary to keep it liquid at normal temperatures. By about 1946 propane was being carried in barges under pressure. With increasing demand for these products a cheaper means of carriage and storage was necessary and it soon became clear that, apart from small quantities, it was advantageous to provide refrigerated storage facilities.

A ship of this type would have a double skin hull construction in way of the cargo space and a double bottom from the fore peak bulkhead to the after peak bulkhead. The longitudinal and transverse bulkheads surrounding the cargo tanks, the tank top plating and deck plating in way of the cargo spaces would be special steel to withstand the low temperatures. The bottom and vertical walls between the cargo tanks and hull would be well insulated. Such a ship would be able to transport propane, propylene, butane and ammonia, the carrying temperatures being as follows:

| Ammonia | $-33°C$ | Propylene | $-45°C$ |
| Propane | $-45°C$ | Butane | $-5°C$ |

The midship section of such a ship showing the construction of the cargo tank is shown in Figure 4-10 by courtesy of the "Shipping World and Shipbuilder".

Methane, unlike the gases already mentioned, cannot be liquefied at normal temperatures irrespective of the pressure applied. The first vessel to transport methane was the *Methane Pioneer,* a converted general cargo ship, which entered service in 1958. This ship was fitted with aluminium tanks and insulation specially designed to hold liquid methane at atmospheric pressure. It having been shown that natural gas could be profitably traded overseas the next step was ships specifically constructed to carry LPG and this was done in 1964. These were the *Methane Princess* and the *Methane Progress.* The hull is of double construction over the entire length of the cargo space. The cargo space is divided into three insulated holds separated by cofferdams and each cargo hold contains three aluminium cargo tanks which conform more or less to the shape of the hull. At the cargo temperature of $-160°C$ normal shipbuilding steel is brittle and thus liable to fracture under stress. Adequate insulation is thus of the greatest importance.

In the transport of liquefied gases at very low temperatures mild steel is quite unsuitable because of its low notch toughness at the low temperature concerned. This problem is resolved in two ways by (1) using aluminium alloys and (2) employing special steels which have good properties at low temperatures.

1) Aluminium magnesium alloys show excellent properties at low temperatures. They are not subject to brittle fracture and consequently have been successfully employed in the construction of the tanks in liquefied gas carriers.
2) As a result of extensive research in the metallurgical field special steels have been developed having good properties at low temperatures. These steels embody nickel as the principal alloying element.

Intermediate between methane and propane is ethylene with a

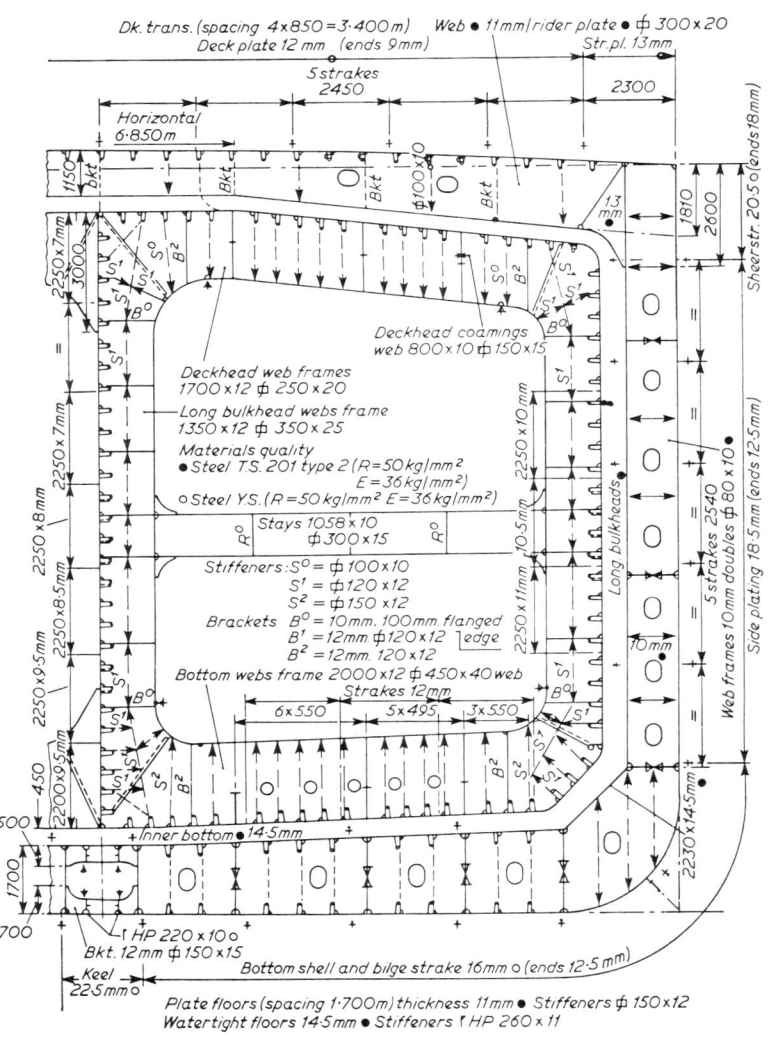

Fig. 4–10.

boiling point of −104°C. Ethylene has to be carried either fully refrigerated or partially refrigerated, as at normal temperature, the pressure to keep the gas liquid becomes too high to be economical.

The first vessel in the world to be built specially for the transport of ethylene (1966) was the *Teviot*. The liquefied gas is contained in a free-standing all-welded aluminium-magnesium alloy tank capable of expansion and contraction in all directions. Part of the cargo tank in the form of a dome pierces the steel trunk top. The void space between the cargo hold and the aluminium tank is filled with nitrogen gas for safety purposes and is topped up from pressure bottles located in a gas-tight compartment in the poop. All the internal surfaces of the cargo hold are heavily insulated to form a secondary barrier protection for the steel hull.

Ships are in service to transport either as a main or alternative cargo the following:

Chlorine; ethylene oxide; butadine; vinyl chloride.

Chlorine: Is extremely toxic and can be carried satisfactorily when dry; if allowed to come into contact with moisture then acute corrosion can be expected.

Ethylene Oxide: This forms explosive mixture with air and in the presence of certain catalysts it can polymerize heat that may cause detonation.

Other cargoes are liquid oxygen and liquid hydrogen. The liquid oxygen is carried at $-183°C$. The liquid hydrogen boils at $-253°C$ and is discharged by overpressure only.

Chemical Carriers

Quite a number of cargoes which today are classified as "chemicals" have been carried at sea for many years and in general these whether solid or liquid were transported in casks, kegs or barrels. With expanding industries demanding very large quantities of specific raw materials the need arose for ships to carry such cargoes in bulk. These ships for the simultaneous carriage of cargoes which, from a quality control aspect, have complete segregation from each other in so far as the pumping and tank arrangements are concerned have been variously described as product carriers or parcels tankers and indeed have been referred to as "floating drug stores".

Many of the chemicals carried are highly corrosive, poisonous and volatile. Consequently, such cargoes have to be handled with care and in the ships which carry them there are problems of construction, operation and maintenance. The liquid chemical carrier belongs to a group which inevitably will keep increasing as the world chemical industry expands.

In May 1969 a Sub-Committee of IMCO issued a code of practice on the carriage of dangerous chemicals in bulk. This code was submitted to the IMCO Assembly in 1971.

The IMCO code visualizes three types of ships—types I, II, III—designed to minimize in varying degrees the effects of collision or grounding, the most hazardous chemicals being allocated to type I. The requirements for each ship type are related to (a) cargo containment and (b) survival capability.

Type I requires a double bottom in way of cargo tanks used for the named chemicals and these cargo tanks are to be not closer than B/5 from the ship's sides measured at the load water line. Only a small number of chemicals are designated for such carriage. The tanks outboard of B/5 may be used for less hazardous cargoes such as allowed for Type III ships.

Type II is to be similarly designed as far as the double bottom is concerned but the relevant cargoes can be carried to 760 mm from the ship's side, the side cofferdams taking account of minor side damage.

Type III for the remaining cargoes is an ordinary tanker in so far as tank arrangement is concerned.

The code also has recommendations to make about the location of accommodation, ventilation of pump rooms, pumping arrangements, tank venting, tank gauging etc., all with the intention of minimizing health hazards to the crew and the environment.

Lloyd's Register of Shipping issued as part of Notice No. 3 to the 1971 Rules: "Provisional rules for the classification of tankers intended for the carriage of liquid chemicals in bulk."

These deal principally with the cargo containment aspect so far as ship arrangement is concerned as the damage survival aspect is at present the responsibility of National Authorities.

CHAPTER 5
The Ship Girder and Structural Details

A ship is often compared with a loaded beam or girder and in some respects the comparison is rational. With steel girders in constructional engineering close approximations can be made as to the nature and extent of the severest stresses which could be induced in the structure. With a ship, due to the very different conditions of service, the problem is much more complex. The varying and sudden stresses experienced by ships in a seaway when rolling and pitching in light, ballast or load condition are such as to preclude any completely accurate mathematical treatment. The conditions present probably the most formidable and complex of all structural engineering problems.

However, as is shown in Chapter 14, account can be taken of the various conditions of ballasting and loading and a degree of accuracy obtained sufficient for comparative purposes.

The modern ship is, in general terms, made up of steel plating, sections and built-up girders so connected as to provide adequate strength in all parts to withstand the forces acting on the ship under all conditions of service.

The forces acting on a ship may be static or dynamic. The static forces are due to the differences in the weight and buoyancy which occur throughout the ship. The dynamic forces are caused by the motions of the ship at sea and the actions of the wind and waves.

These forces create:
1) Longitudinal stresses
2) Transverse stresses
3) Local stresses

The greatest stresses set up in the ship as a whole are due to the distribution of loads along the ship, causing longitudinal bending.

Longitudinal Stresses

The simplest way to consider this is to imagine the vessel floating in equilibrium in still water. The forces are two in number, the weight of the ship and all that it carries acting vertically downwards and the vertical component of the hydrostatic pressure. The total buoyant force is equal to the total weight but the distribution of the weight and buoyancy along the length of the ship are not similar.

The buoyancy per unit length is given by

$$A \times \text{density of fluid}$$

where A is the immersed cross-sectional area of the ship at any given position in the length of the ship. If these values at different positions along the length of the ship are plotted on a base representing the ship's length then a curve of buoyancy is formed. Such a curve for a ship in still water is shown in Figure 5-1. The area of this curve gives

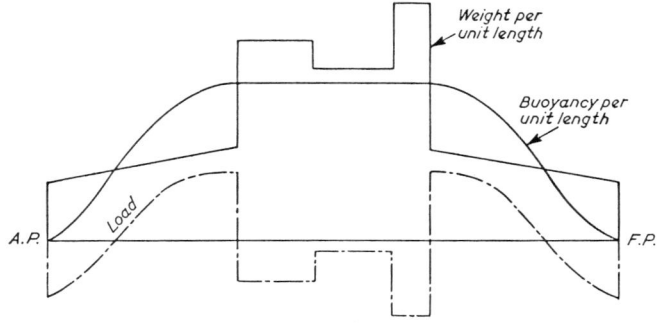

FIG. 5-1—*Curve of buoyancy.*

the total buoyancy. The weight per unit length depends upon various factors such as the position of the items that make up the whole, the density of the cargo, fuel etc. and their disposition. Such a curve is also shown in Figure 5-1. The difference between the weight and buoyancy at any point is the load at that point. In some cases the load is an excess of weight over buoyancy and in others an excess of buoyancy over weight. The load curve is shown in the diagram.

Since the total weight must be equal to the total buoyancy the area of the load diagram above the base line must be equal to the area below the base line. Because of this differing distribution of weight and buoyancy or unequal loading along the length of the ship, shearing forces are set up and a bending moment which causes the ship to bend in a longitudinal vertical plane like a beam. The shearing force curve is obtained by integrating the load curve. The bending moment curve is obtained by integrating the shearing force curve. Further details are given in Chapter 14.

Depending upon the direction in which the bending moment acts the ship will hog or sag. If the buoyancy amidships exceeds the weight, the ship will hog, as a beam supported at mid-length and loaded at the ends. Figure 5-2.

If the weight amidships exceeds the buoyancy the ship will sag, as a beam supported at the ends and loaded at mid-length. Figure 5-3.

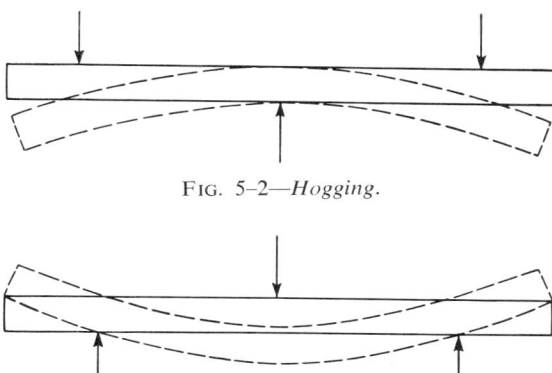

FIG. 5-2—*Hogging.*

FIG. 5-3—*Sagging.*

When a ship is moving amongst waves the distribution of weight remains unchanged but as the surface of the water is no longer flat the distribution of the buoyancy along the length of the ship is altered. Figure 5-4 shows the ship poised on a wave whose length is equal to that of the ship. When the wave crest is amidships the buoyancy amidships is increased while at the ends it is reduced. This tends to cause the ship to hog.

In Figure 5-5 the wave trough is amidships. Here the buoyancy amidships is reduced while at the ends it is increased, causing the vessel to sag.

A more detailed discussion of the foregoing is given in Chapter 14.

Transverse Stresses

A transverse section of a ship is subject to static pressure due to the surrounding water as well as internal loading due to the weight of the structure, cargo etc. The effects of these are not as great as those in the longitudinal direction. If the ship has adequate strength in the longitudinal direction then the normal methods of construction ensure satisfactory conditions in the transverse direction.

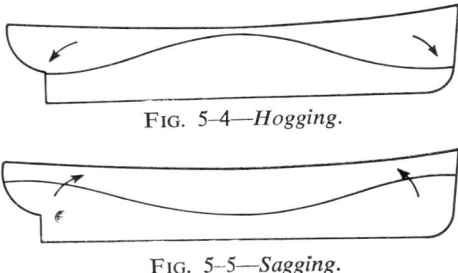

FIG. 5-4—*Hogging.*

FIG. 5-5—*Sagging.*

The parts of the structure which resist transverse stresses are:
a) Transverse bulkheads
b) Floors in the double bottom
c) Brackets between deck beams and side frames together with brackets between side frames and tank top plating or margin plate
d) Pillars in holds and 'tween decks.

Dynamic Forces

The dynamic effects arise from the motion of the ship itself. A ship among waves has three linear motions:
1) Vertical movement—*Heaving*
2) Transverse movement—*Swaying*
3) Fore and aft movement—*Surging*

and three rotational motions:
1) *Rolling* about a longitudinal axis
2) *Pitching* about a transverse axis
3) *Yawing* about a vertical axis.

When the ship motions are large particularly in pitching and heaving, considerable dynamic forces can be created in the structure.

Panting

As waves pass along the ship they cause fluctuations in water pressure which tend to create in and out movements of the shell plating. This is particularly the case at the fore end. The rules of the classification societies require extra stiffening, at the ends of the ship, in the form of beams, brackets, stringer plates etc. in order to reduce the possibility of damage: this in and out movement is called panting.

Slamming or Pounding

In heavy weather when the ship is heaving and pitching the fore end emerges from the water and re-enters with a slamming effect which is called pounding. The classification rules require extra stiffening at the fore end to reduce the possibility of damage.

Local Stresses

These are created by such items as:
a) Heavy concentrated loads like engines, boilers etc.
b) Deck cargo such as timber
c) Hull vibration
d) Ship resting on blocks in a dry dock.

The Ship Girder

Since the ship is capable of bending in a longitudinal vertical plane it follows that there must be material in the ship's structure to resist this bending. The material distributed over a considerable portion of the ship's length which contributes to longitudinal strength

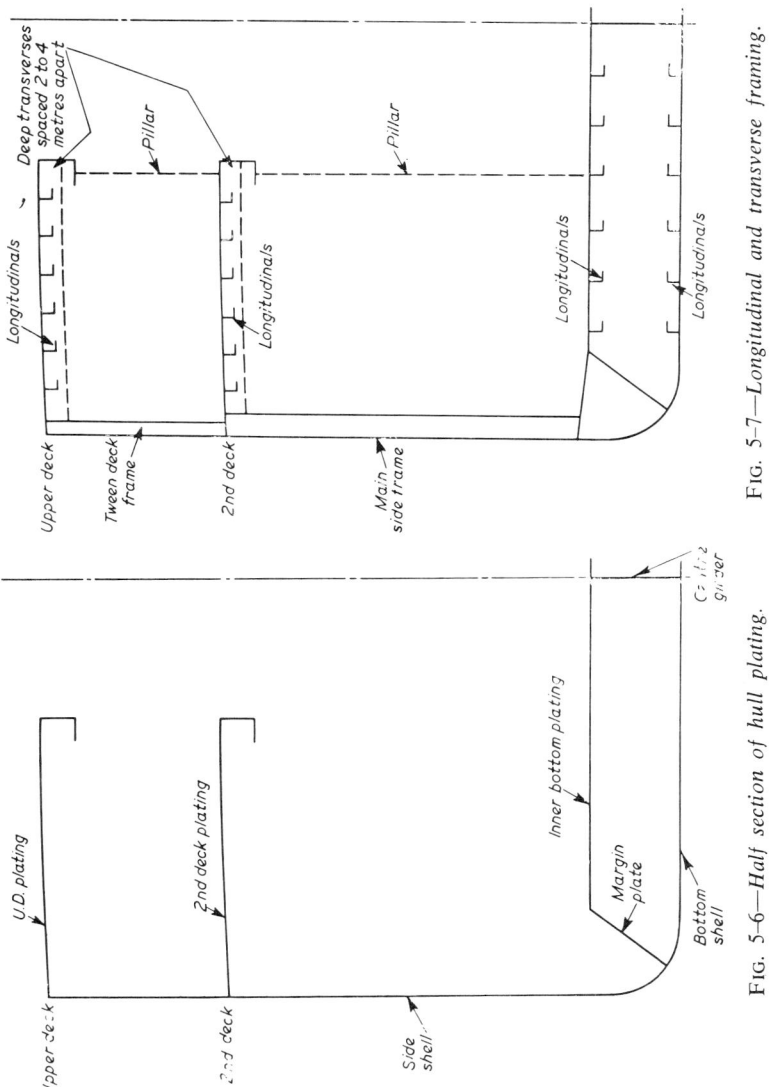

FIG. 5-7—Longitudinal and transverse framing.

FIG. 5-6—Half section of hull plating.

consists of the side and bottom shell plating, inner bottom plating, centre girders and decks. Figure 5-6 is an outline section showing these items. It is usual, as far as the decks are concerned, to consider only the material abreast the openings of the hatches and machinery casings.

This longitudinal material forms a box girder which has very large dimensions in relation to its thickness. Hence the plating must be stiffened so as to be able to withstand compressive loads. This is achieved in the dry cargo ship by a form of construction in which the sides of the ship are stiffened transversely whilst the decks and bottom are stiffened longitudinally as shown in Figure 5–7. These longitudinals also contribute to the general longitudinal strength of the structure. Moreover the longitudinals have to sustain cargo and water pressure loads and to reduce their scantlings they are supported at positions other than at the bulkheads by deep transverse beams in the decks and transverse plate floors in the bottom. These beams and plate floors are spaced from 2 to 4 metres apart. The decks are supported by pillars as shown in Figure 5–7.

FIG. 5–8—*Hoisting 730 tonne prefabricated bow section*

Welding

The use of welding in place of riveting as the means of connecting structural elements has not only influenced their form and size but also the mode of their assembly. It has indeed brought about the now universally adopted prefabrication technique. Some of the prefabricated units are very large as instanced by the fore end unit of a giant tanker built by Harland and Wolff at their Belfast shipyard. This bow unit of 730 tonnes was hoisted into position by what was claimed to be the largest crane in the world. Figure 5-8.

The modern method of welding is by fusion weld where there is actual melting of the material being joined although the weld metal is composed largely of added material having the same characteristics as the metal being joined. The application of welding to shipbuilding is almost entirely fusion welding in the form of metallic-arc welding. This is the most versatile welding process available. A metal electrode, a coated steel rod, is held in a special holder through which an electric current is passed and which strikes an arc between the end of the electrode and the work-piece. The heat generated melts both the electrode tip and a small portion of the work-piece. The metal and flux from the electrode are transferred to the work-piece forming a molten pool of metal protected from the atmosphere by a covering of molten slag formed from the flux. The flux also provides an inert gas which steadies the arc and protects it from the atmosphere. Many of the electrodes are of the "all positional" type which means they can be used either in the flat, vertical, overhead or inclined position.

The main advantages of welding over riveting are:

a) appreciable saving in steel mass
b) economy of construction
c) ease of fabrication
d) continuity of strength
e) watertightness without the need for caulking.

There are, of course, other aspects in that the efficiency of the joints depends upon the welder; the rapid heating and cooling tends to distort the plates; a faulty weld is difficult to detect.

The checking of welds is carried out by radiography in the form of X-ray and gamma ray techniques, the former being the most common. Radiographs are taken of important butt welds by passing the rays through the plate on to a photographic plate. Any differences in the density of the plate allow greater exposure and may readily be seen when developed.

The use of ultrasonics is being steadily established as a system of checking. A high frequency electric current causes a quartz crystal to vibrate at a high pitch. The vibrations are transmitted directly through the material being tested. If the material is homogeneous, the vibration is reflected from the opposite surface, converted to an electrical impulse and indicated on an oscilloscope. Any fault in the material, no matter how small, will cause an intermediate reflection which may be noted on the screen.

Rolled Steel Sections

The connecting of plates and their stiffening when riveting was in vogue required various forms of material. These forms were termed sections and are known as "rolled steel sections". They consist of ordinary angles, bulb angles, bulb plates, channels, as shown in Figure 5–9.

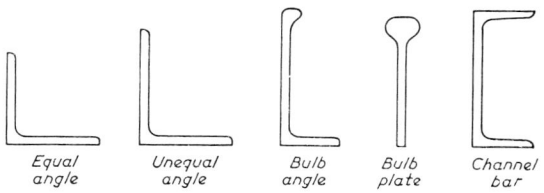

FIG. 5–9—*Rolled steel sections.*

In welded ships, connecting angles are no longer required but unequal angles are used by toe-welding them to the plates and thus forming excellent stiffeners as in Figure 5–10. For welded work the flange of the bulb angle is not required for connecting purposes thus saving weight and a bulb plate has been developed with a heavier bulb as in Figure 5–10.

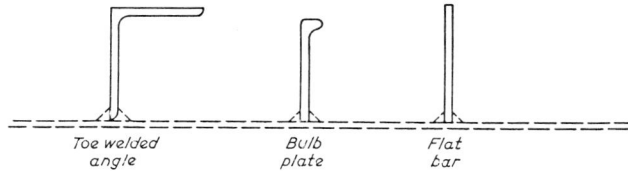

FIG. 5–10—*Welding of angle to plate.*

Welded Connexions

The design of welded joints is an extremely important part of welding procedure. Correctly applied it helps to control distortion, reduce shrinkage, produce good workmanship and sound welds economically. The two basic types of welded connexions are the butt and fillet.

The most efficient method of joining two plates in the same plane is by means of a butt weld. In this way the two plates become one continuous member. A square edge butt may be used for plates under 1 cm thick. Above this thickness it is rather difficult to secure adequate penetration and it becomes necessary to adopt single Vee or double Vee butts. These are shown in Figure 5-11 as is the overlap joint. The latter is used only for joints of relatively minor importance.

Fillet welds are employed between plate surfaces which normally meet at right angles, Figure 5-11. The size of a fillet weld is specified

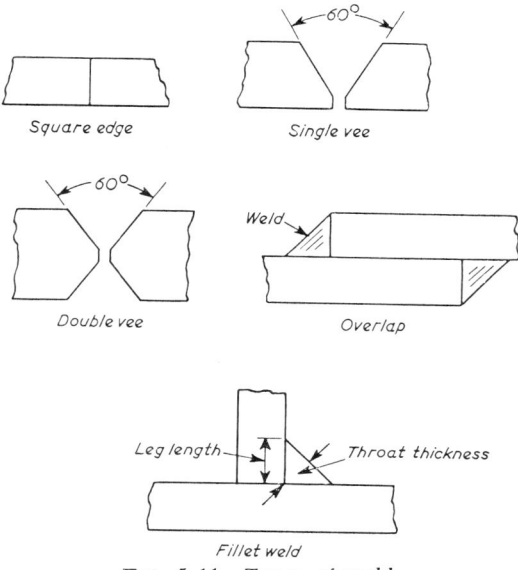

FIG. 5-11—*Types of weld.*

by its leg length and throat thickness, the latter being at least 70 per cent of the leg length. The welds may be continuous on one or both sides of the member or may be intermittent. Continuous welds are used when the joint must be water tight and for strength members.

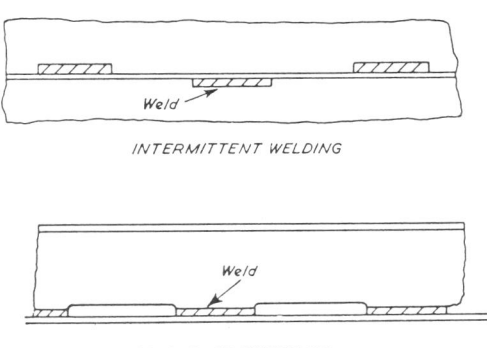

FIG. 5-12—*Welding of stiffeners, frames and beams.*

Stiffeners, frames and beams may be connected to the plating by intermittent welding. Where the rate of corrosion is high it is essential to have continuous welding or to scallop the section as in Figure 5–12. The latter method has the advantage of reducing the weight of the structure and at the same time improving the drainage.

General

The position of joints and the sequence of fabrication should be such as to give maximum accessibility to the operator and to permit as much welding as possible in the flat position.

Structural discontinuities should, as far as possible, be avoided so as to reduce the possibility of cracks which may occur at points of stress concentration caused by discontinuities.

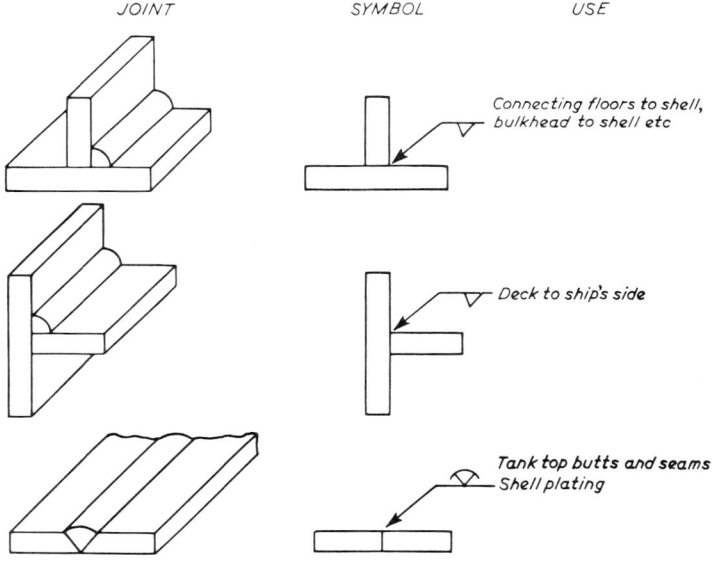

FIG. 5–12a(1)—*Welding symbols.*

Welding Symbols

Figure 12A illustrates some types of weld employed and the symbols used on drawings to indicate them.

A comprehensive list of symbols is given in BSI 499.

Safety in Welding

Fires. In the process of welding fires are always a risk. Welding should be avoided in way of combustible material.

Gases. Gases such as oxygen, carbon dioxide, hydrogen and propane, which can form highly inflammable or explosive mixtures are

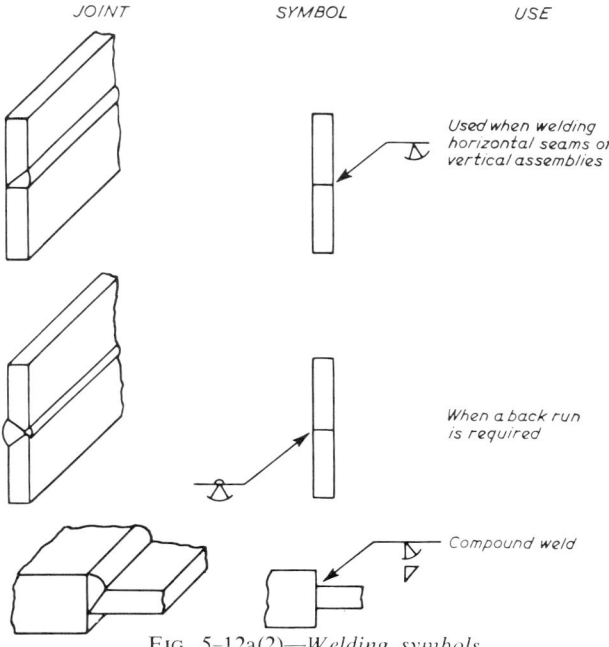

Fig. 5-12a(2)—*Welding symbols.*

used for welding and heating processes and obviously care must be taken both in their storage and use.

Electric Shock. The insulation of cables and welding equipment should be kept in perfect condition in order to reduce the dangers. Efficient earthing arrangements are imperative.

Burns. Protective clothing must be worn to protect the operator from burns by contact with hot metal or by exposure to the rays emitted by the welding arc.

Eye Injuries. To prevent the ultra-violet and infra-red rays reaching the eyes and to reduce intensity of the visible light rays, it is essential to have glass filters of suitable properties between the eyes and the arc.

Offensive Fumes. Adequate ventilation must be provided to remove the fumes produced when welding.

Elements of Structure

Prefabrication in large units makes it impossible to present the elements of structure in the traditional form of erection—keel, framing, longitudinals, bulkheads, decks etc. Some details, however, are given of the main components of the structure.

Figure 5–13 gives the midship section of a cargo ship showing the general disposition of the steel material.

Keel and Duct Keel (FIG. 5–14).

The keel is frequently referred to as the backbone of the hull

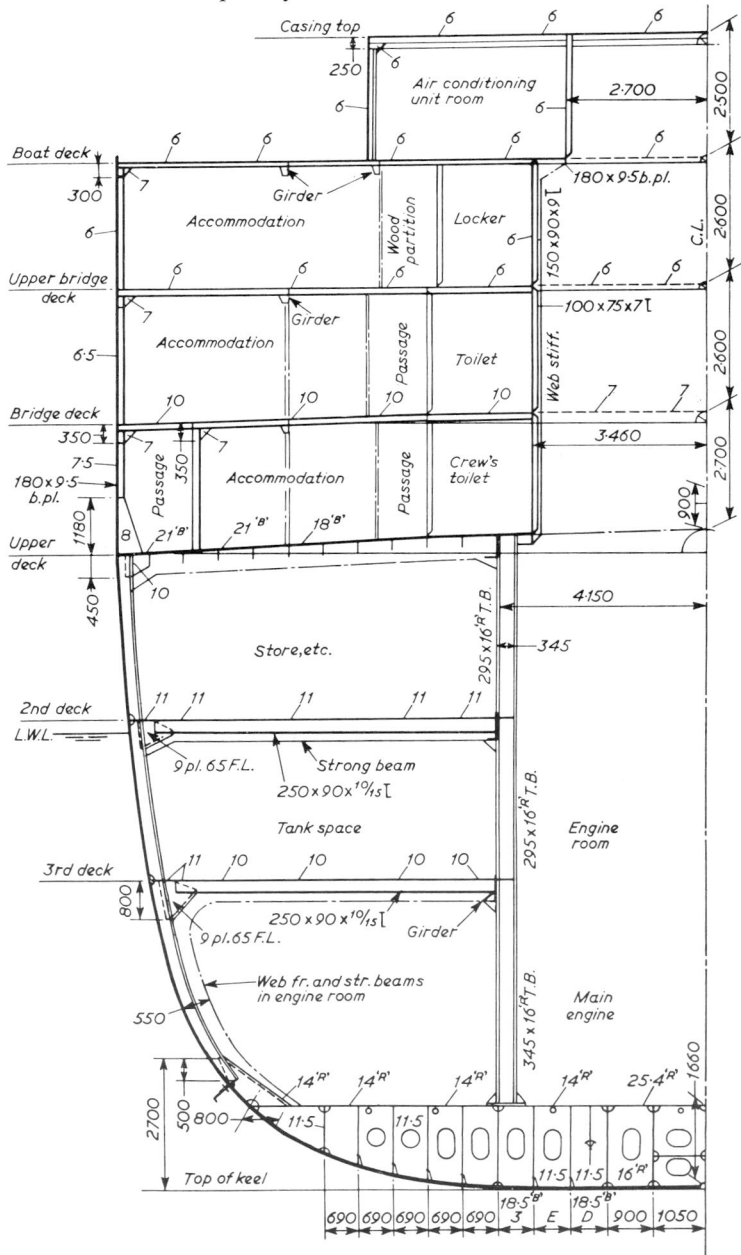

THE SHIP GIRDER AND STRUCTURAL DETAILS 55

because of its important contribution to resisting bending action. It consists of the centre girder, the middle line strake of the inner bottom

FIG. 5–13—*Midship section of cargo ship.*

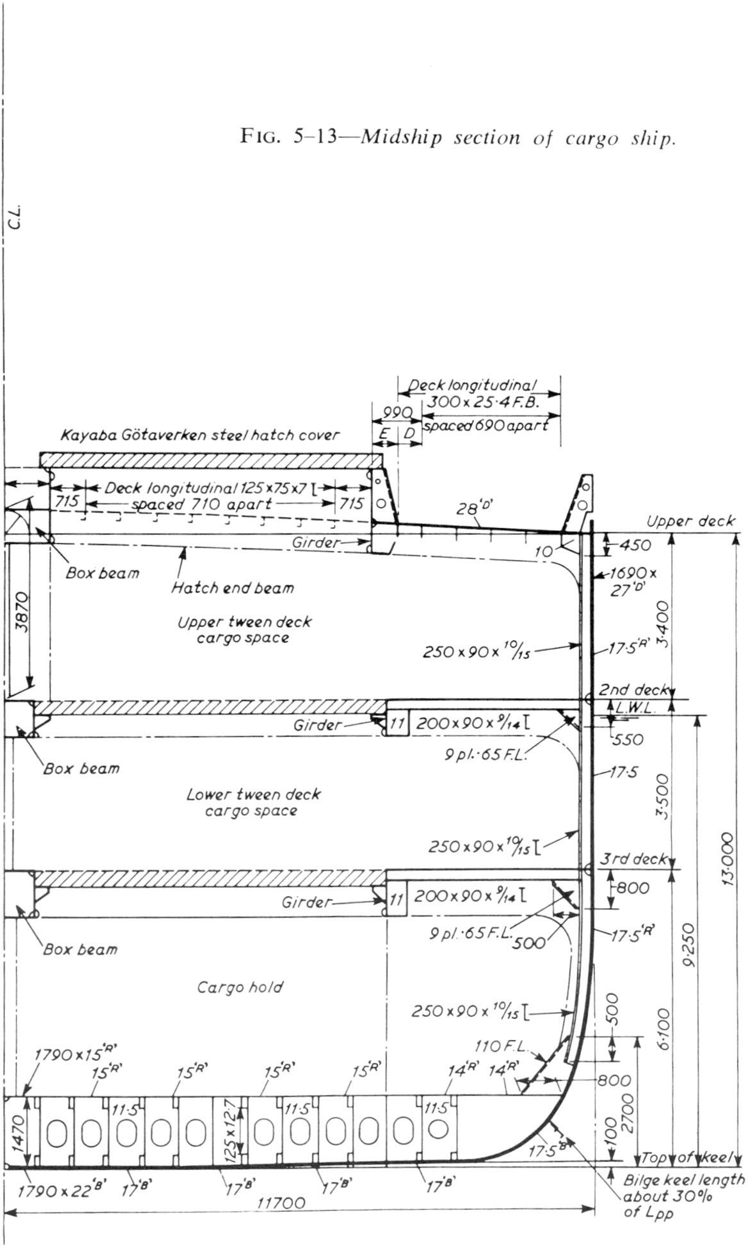

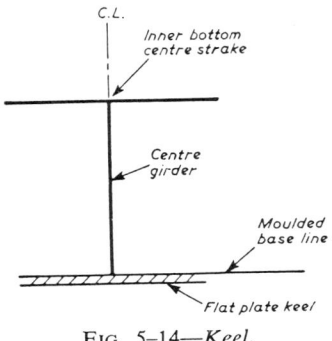

Fig. 5-14—*Keel.*   Fig. 5-15—*Duct keel.*

plating and the flat plate keel or middle line strake of the outer bottom plating. The centre girder is generally watertight within the length of the double bottom to maintain strength and sub-divide the fuel and water tanks.

Figure 5-15 shows a duct keel arrangement which has the advantage that the fore and aft piping from the various compartments of the double bottom can be taken along this space and are thus more readily accessible.

Double Bottom

The double bottom about 1 to $1\frac{1}{2}$ metres in depth, creates a structure which provides a measure of protection in the event of damage to the outer shell, without flooding the holds or machinery space, so long as the inner bottom remains intact. A very important advantage is in the provision of compartments for the carriage of oil fuel, fresh and feed water and water ballast. A cofferdam must be fitted between between a fuel tank and a fresh water tank to prevent contamination. The centre girder which is unpierced acts as a division between tanks thus making the tanks in pairs, one port and one starboard. Additional longitudinal side girders are fitted depending upon the breadth of the ship but these are neither continuous nor watertight.

Figure 5-16 shows a double bottom framed longitudinally. This type of construction gives an accessible double bottom and also considerable longitudinal strength. Between the widely spaced floors transverse brackets are fitted at each frame at the margin plate and about 1 m apart at the centre line.

Floors. A floor plate is a transverse vertical plate that runs across the bottom of the ship from the centre girder to the bilge. Watertight or oiltight floors are used to divide the double bottom space into suitable tanks.

THE SHIP GIRDER AND STRUCTURAL DETAILS 57

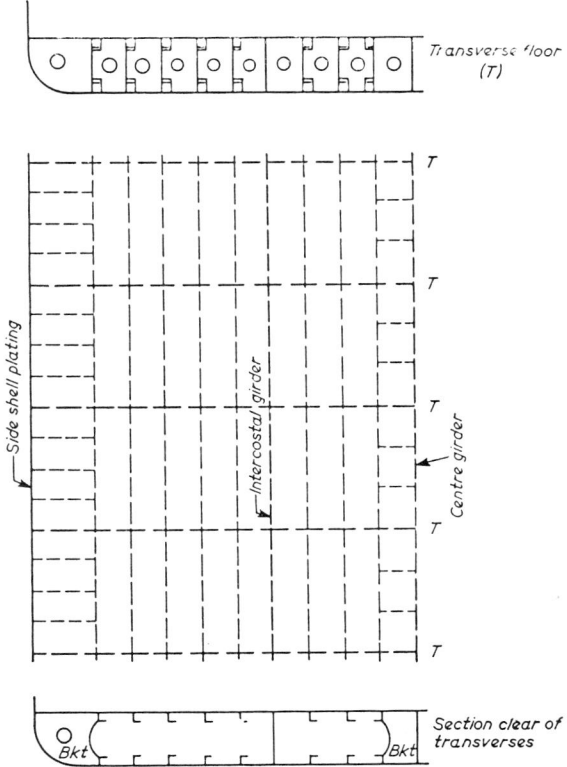

Fig. 5-16—*Double bottom.*

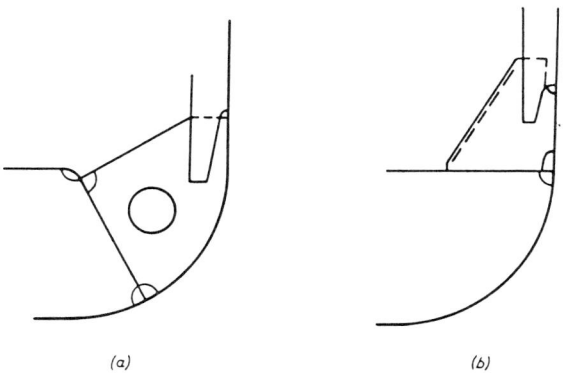

Fig. 5-17—(a) *Sloping Margin.*
(b) *Flat Margin.*

Intercostal Girders are parallel to the centre girder and fitted between the floors.

The inner bottom plating is arranged in longitudinal strakes. The middle line strake is somewhat thicker than the remainder as it constitutes part of the so-called back-bone of the hull. Access to the double bottom is provided by means of man-holes cut in the inner bottom or tank top plating. Figure 5–17 (a) shows the sloping margin which is connected to the shell plating by a bracket and (b) the horizontal margin now extensively adopted. Where a margin is fitted forming a bilge well as in (a) the branch bilge suctions are taken through the wing brackets. In a level tank top as in (b) a bilge well must be provided port and starboard.

Air pipes. These are fitted to double bottom tanks at the highest part of the tank or at the opposite end to that at which filling takes place. They are used to release the air from the tanks when being filled and the air pipes are led to the open. Satisfactory means permanently attached must be provided for closing the openings of the air pipes.

Sounding Pipes. These are used to determine the depth of liquid in the tanks. They should as far as possible be straight but where this is not possible the curvature should be such as to allow the ready passage of the sounding rod or chain. Striking pads should be placed under the lower end of sounding pipes or the ends of the pipes closed and perforated with small holes to admit the liquid.

Testing. All double bottom tanks prior to launching and at subsequent special surveys have to undergo pressure tests. The compartments are filled with water and then pressed to the greatest head likely to be experienced in practice, i.e. height of overflow or air pipe.

Double Bottom in the Machinery Space. Engine Seating

The structural strength of the ship in way of the propelling machinery space demands special attention. The main units of the propelling machinery are supported on seatings which generally take the form of thick plates. These seatings are in turn supported by special longitudinal and transverse members and bearers in which the welding is continuous; intermittent welding is not adopted.

As well as supporting the weight of the machinery the bearers must be able to withstand the stresses created by the propeller thrust, vibration and motion in a seaway. Moreover, they must be sufficiently rigid to ensure permanent alignment when the ship is at sea on full power.

The height of the seatings relative to the keel depends on the position of the shaft as it enters the engine room and this depends on the diameter of the propeller. These factors demand a built up engine seat or an increase in the depth of the double bottom in way of the engine room where the engine bedplate is bolted direct to heavy plates incorporated in the tank top, the bolts pentrating into cofferdams. Where the double bottom is increased in depth above the normal, continuity of strength is maintained by the gradual sloping of the tank top and intercostal girders between the machinery space and the adjacent compartments as in Figure 5–18. All transverse floors

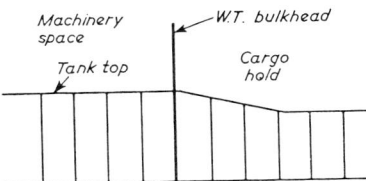

FIG. 5–18—*Double bottom depth.*

at every frame in the machinery space are of the type known as solid, and additional fore and aft strength is obtained in the form of intercostal and continuous plate girders. Particular attention is given to the supports for the thrust block as this takes the full thrust of the propeller. These supports must be able to cope with the fluctuations in propeller thrust which arise due to rolling and pitching in rough weather.

Boiler Seatings

A boiler either cylindrical or water-tube is a heavy item but the question of its support is not so involved as in the case of the engine. The double bottom has floor plates at every frame and the thickness of these and the longitudinal girders is increased. For the cylindrical boiler vertical cradle stools are provided, these being cut to fit the curvature of the boiler shell. At least two such stools are required as well as some form of longitudinal bracing to prevent the stools from tripping. A collision chock is fitted at the fore end to prevent the boiler from moving forward and thus disrupting steam pipes. In the case of the water-tube type of boiler most of the boiler weight is taken by strong supports erected under the feet of the water drum. The construction of the boiler and its seating arrangements are such that no other structural connexions are required to steady it against rolling and pitching. The feet at one end of the boiler are fixed and the remainder are arranged to slide to allow for expansion. Grease grooves are cut in the bases of sliding feet and the holding down bolts work in elongated holes.

Shaft Tunnel

When the machinery space is not at the aft end of the ship and is separated from the after peak by one or more cargo holds, the main shafting must be carried through the holds in a tunnel. The tunnel prevents contact with cargo and gives access to the shafting at all times for maintenance and repair. The tunnel is watertight and extends from the after bulkhead of the machinery space to the after peak bulkhead. The top of the tunnel is generally circular as in Figure 5–19 except at the extreme aft end where it is more convenient for

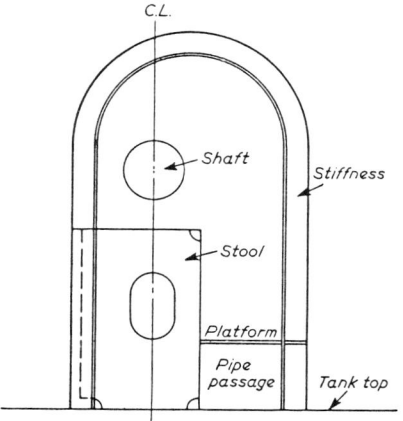

Fig. 5–19—*Shaft tunnel.*

it to be flat. As a passageway is not necessary on both sides of the shaft the tunnel is constructed off the middle line of the ship allowing a passage on one side. The tunnel stiffeners are fitted inside the tunnel except in the way of an insulated hold where they are outside to facilitate the arrangement of the insulation. In way of hatches the tunnel top plating is increased in thickness unless wood sheathing is fitted to prevent damage. The tunnel is in general about $1\frac{3}{4}$ m in breadth and $2\frac{3}{4}$ m in height.

The shaft tunnel is used as a pipe tunnel, the pipes being carried along the tank top with a walking platform about $\frac{1}{2}$ metre above the tank top plating. A watertight door is fitted in the after bulkhead of the machinery space giving access to the tunnel from the machinery space. At the aft end of the tunnel a watertight escape trunk is fitted and this extends to the deck above the deep load line.

The full length of the propeller shafting is made up of three or more lengths; the tail shaft which extends from the propeller through the stern tube and stern gland; one or more lengths—intermediate shafts—extending from the stern gland to the thrust shaft of the thrust block. The shaft is supported at intervals by bearings which are fitted on shaft stools. These stools are of necessity strong and rigid in order to maintain correct alignment of the shafting.

Framing

The framework which stiffens the hull of a ship is in effect the combination of two systems of framing—the longitudinal and transverse systems. As stated previously the general practice today for the dry cargo ship is a form of construction in which the sides of the ship are stiffened transversely whilst the decks and bottom are stiffened longitudinally. Figure 5–7. The types of section in common use for frames are the ordinary angle with toe welded to the shell plating, the bulb plate, and the flat bar as in Figure 5–10. The side framing acts as stiffeners holding the side shell against external water pressure. Transverse side framing is preferred since the deep transverses required to support longitudinal framing in both holds and 'tween decks have a great disadvantage in dry cargo ships in that they interfere with cargo stowage. The transverse side frames are spaced up to about 1 metre apart. The spacing depends upon the length of the ship.

The longitudinal system of framing has also been used for the sides of the ship as well as the decks and bottom. In 1908 the late Sir Joseph Isherwood patented a longitudinal system of ship construction

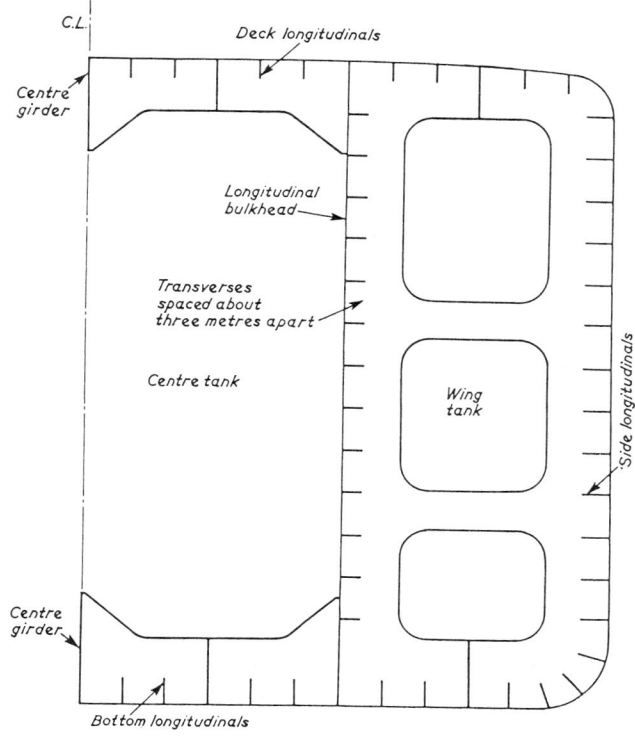

FIG. 5–20—*Longitudinal system of framing—tanker.*

and many tankers were built to its design. The system consisted of a series of closely spaced longitudinal girders with widely spaced transverse webs. Later in the development of the tanker the combined system of longitudinals in the bottom and deck with transverse side framing was adopted. However, a number of the large tankers today have been constructed on the complete longitudinal framing system. The midship section of such a tanker is shown in Figure 5–20.

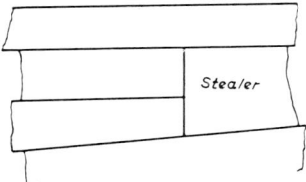

FIG. 5–21—"*Stealer*" *plates at forward and after ends.*

Shell Plating

The shell plating in addition to acting as a watertight skin contributes largely to the structural strength. It is, indeed, the principal longitudinal strength member of the hull girder. The mass of the shell plating forms a fair proportion of the total mass of the steel hull—in the region of 25 per cent. The shell plating is arranged in a series of fore and aft strakes and the strake of side plating at the deck is known as the sheerstrake. The thickest strakes are the flat plate keel and the sheerstake. The thickness is decreased towards the ends. As the girth of the ship is considerably less at the ends of the vessel than at amidships the number of strakes is reduced—to avoid very narrow strakes—by the introduction of "stealers". A stealer plate replaces two plates in adjacent strakes and involves these two plates ending at the same place. Figure 5–21.

The bottom shell is strengthened forward to resist pounding and to resist panting the hull is strengthened in way of the after peak tank and at the fore end. A rounded gunwale is now quite common in tankers and dry cargo ships as shown in Figure 5–22(a). It is an improvement on that of (b) as it eliminates the source from which a notch-brittle fracture may emanate.

In way of openings in the shell plating, break in the erections and discontinuity of strength the plating is increased in thickness.

Decks

Steel decks have three important functions.
1) they must contribute to longitudinal strength
2) they must be capable of supporting the deck loads that may be imposed on them
3) they must, when closed, satisfactorily complete the watertight hull of the ship.

The deck plating is made up of a series of longitudinal strakes and the strakes that adjoin the ship's sides, called stringer plates, are of great importance as they form the connexion between the deck and the shell. They are generally of greater thickness than the remainder of the deck plating. The connexion of the stringer to the sheerstake is shown in Figure 5-22. At the ends of the ship the thickness of

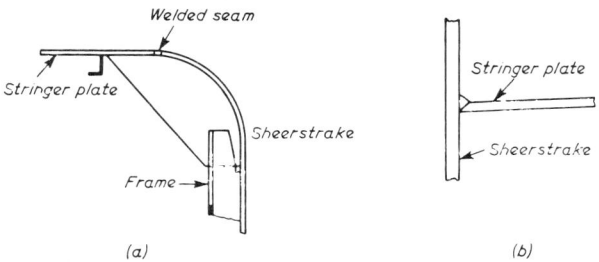

FIG. 5-22—(a) Rounded gunwale.
(b) Alternative arrangement—less efficient.

plating may be gradually reduced in the same way as the shell plating. The plating between the hatches has little effect on longitudinal strength.

The weather deck and the decks above are cambered to assist drainage. The camber is generally parabolic and of the order of 1/50 of the breadth of the ship amidships.

The large openings in the decks for cargo hatchways, in way of machinery casings, pump room entrances etc. create reductions in the cross-sectional area of the deck plating and thus produce a discontinuity of strength. The classification society rules specify a sectional area for a deck and this must be maintained; consequently abreast a deck opening the plating is thicker than normal.

The exposed steel decks over accommodation must be sheathed with wood or some approved deck composition. This acts as heat and sound insulation. It is essential that the deck be adequately protected against corrosion. There are many substitutes for wood sheathing and they have the advantage of being in general, non-inflammable, and requiring little service.

The decks may be supported either by transverse beams in conjunction with longitudinal girders or by longitudinals in conjunction with transverse girders. The latter type of construction is common practice today. The effect of stiffening the deck and bottom by longitudinal members instead of transverse members is to increase very greatly the buckling strength of the plating and it is largely for this reason that this method of construction has been adopted.

When a ship sags, the upper longitudinal members are subjected to compression and the bottom shell plating, tank top etc. to tension.

When a ship hogs, the direction of these forces is reversed. The thickness of deck plating, shell plating and tank top plating is small compared with the ship's depth and although suitable for tension forces is less effective in compression.

The deck longitudinals extend, as far as practicable, along the entire length of the ship outside the line of hatches. They are connected to the transverse bulkheads by brackets and are supported by transverse girders which run across the full width of the ship or in way of hatches from the ship side to the hatch.

If cargo is suspended from a deck, as in the case of meat hanging from the deck overhead, then additional strengthening must be introduced in order to carry the suspended load in addition to the normal load on the deck.

Watertight Bulkheads

The primary purpose of transverse watertight bulkheads is to divide the ship into a series of watertight compartments, but since they have very considerable transverse rigidity they contribute largely to the prevention of transverse deformation of the structure. Dividing the ship into watertight compartments means that in the event of damage admitting water to the vessel this water will be confined to the damaged region.

The number of bulkheads in a cargo ship depends upon the length of the ship and the position of the machinery space. For a ship with the propelling machinery amidships the minimum number of bulkheads is four, viz., the collision bulkhead, the after peak bulkhead and a bulkhead at each end of the machinery space. Each ship must have a collision bulkhead not less than 5 per cent of the ship's length from the fore part of the stem at the waterline. This bulkhead must extend to the uppermost continuous deck. The stern tube must be enclosed in a watertight compartment formed by the sternframe and the after peak bulkhead which may terminate at the first watertight deck above the waterline. It should be noted also that the propeller shafting is enclosed in a watertight tunnel. With a bulkhead at each end of the machinery space there is in the cargo ship with the minimum number—four—of bulkheads a certain measure of subdivision. If the machinery is aft the after peak forms the after boundary of the space. Additional bulkheads are required by classification rules, the number depending on the length of the ship.

For a passenger ship the internal subdivision is laid down by international convention. Passenger ships are defined in the Merchant Shipping Acts as those carrying more than 12 passengers. The degree of subdivision into watertight compartments varies according to the nature of the service of the ship and its length on the load waterline. The highest standard of subdivision is applied to a ship primarily engaged in the carrying of passengers and is gradually reduced in order to suit a ship which comes close to the purely cargo type.

The general standard is that in assumed conditions of flooding, the ship will float at a draught not greater than 76 mm below the deck at which the watertight bulkheads terminate i.e. at the margin or 76 mm below the bulkhead deck. For example in a large passenger ship on foreign going voyages the subdivision is designed to maintain

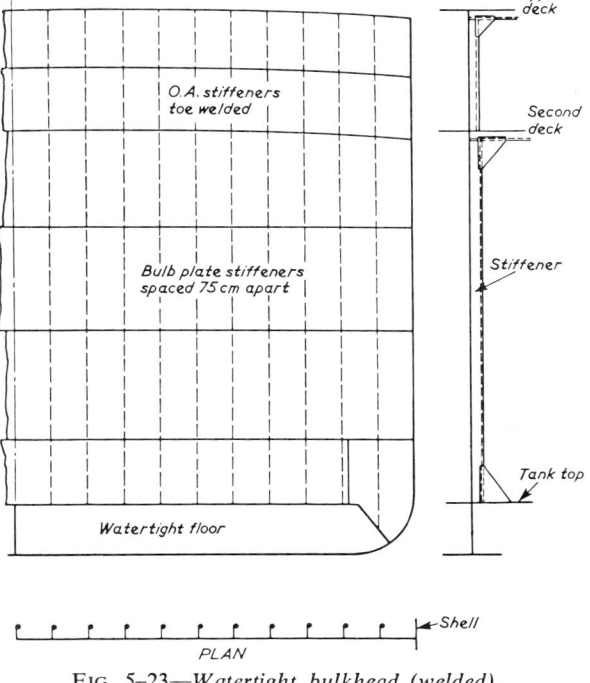

FIG. 5–23—*Watertight bulkhead (welded)*.

a maximum or less draught when three adjacent compartments are laid open to the sea. As stated above, in the lower categories the standard is less stringent, but in general the aim is to ensure that if one compartment is flooded the ship's waterline does not come over the margin line.

Watertight bulkheads are formed by plates which are attached to the shell, deck and tank top. Bulkheads are usually plated horizontally and stiffened vertically. Since water pressure increases with the head and the bulkhead is designed to withstand such loading it is clear that the plating on the lower part of the bulkhead is thicker than that at the top. The plates in way of the bilge are further increased because of corrosion. Figure 5–23.

The bulkheads are supported by vertical stiffeners spaced about $\frac{3}{4}$ m apart. The ends of the stiffeners are, in general, secured by brackets to the tank top and deck. In general the stiffeners are either bulb plates or toe welded angles. The stiffeners are scalloped where they pass over a welded seam so as to avoid welding over the seam.

A watertight floor should be placed in the double bottom directly below a main transverse bulkhead.

If it is found necessary to penetrate a bulkhead it is essential to ensure that the bulkhead remains watertight. The after engine room bulkhead is penetrated by the main shaft, which passes through a watertight gland and by an opening leading to the shaft tunnel: this opening must be fitted with a sliding watertight door.

Deep Tanks

In general it is not practicable to arrange in the double bottom all the ballast that may be required and deep tanks are commonly fitted to meet this need. The deep tank is usually designed to permit dry cargo to be carried and in some cases may carry vegetable oil or oil fuel as a cargo. Deep tanks are also provided for the carriage of oil fuel as bunkers.

Deep tank bulkheads are designed to withstand a head of water up to the top of the overflow pipe. Consequently the strength of the structure is greater than that required for the bulkheads of dry cargo holds. Compared with a hold bulkhead the stiffener spacing is reduced and the size of the stiffeners increased as well as the thickness of the plating.

If a deep tank is to be used as a water ballast tank it should be either completely full or empty while at sea to ensure that there is no movement of water.

The hatches to a deep tank have a dual purpose.
1) they must be watertight or oiltight
2) they must be large enough to permit normal cargoes to be loaded and discharged.

These hatches may be from 3 m to 4 m square suitably stiffened and with means whereby the cover can be secured. They may be hinged or arranged to slide.

Deep Tank for Oil Fuel or Oil Cargo

A deep tank containing oil will have a free surface and as an oil fuel bunker there will be different levels of oil during the voyage. Consequently, there will be a free surface effect causing reduced stability and moreover the surging of the liquid across the tank may cause damage to the structure. The effect on stability is dealt with in Chapter 11. To minimize the surging it is necessary to fit a centreline bulkhead if the tank is the full breadth of the ship. This division may be intact or perforated; generally the latter. The perforations must be such as to prevent a build up of pressure on one side and at the same time exclude surging to any marked extent. Figure 5–24.

Hatchways

A hatchway is a large opening cut in the deck to provide access to the hold spaces. In general there is one hatch per hold or 'tween decks space although with large holds two hatches are sometimes arranged. The length and width of a hatch depend mainly upon the

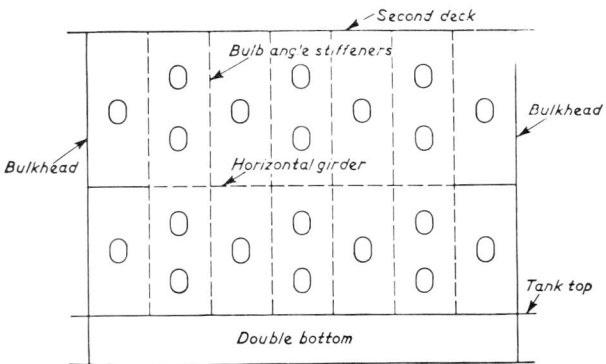

FIG. 5–24—*Elevation of wash bulkhead.*

size of the ship and the type of cargo to be carried. If a restrictive cargo is likely, such as heavy machinery, locomotives, long structural steel etc., then the minimum size is dictated by their requirements. Bulk carriers have long, wide hatches to allow the cargo to fill the extremities without being manually trimmed. There are, however, limiting conditions as very large openings in the deck, particularly the strength deck, reduce the effective cross-sectional area of that important strength member.

It is necessary to protect all weather deck openings against the risk of swamping. Hatchways are framed by the hatch coamings fitted along the sides and ends of the openings; the height of these coamings is important.

Regulations about hatchways are given in the current International Convention on Load Lines and the following are extracts:

For the purpose of the Regulations, two positions of hatchways are defined:

Position 1. Upon exposed freeboard and raised quarter decks and upon exposed superstructure decks situated forward of a point located $\frac{1}{4}$ L from the forward perpendicular.

Position 2. Upon exposed superstructure decks situated abaft $\frac{1}{4}$ L from the forward perpendicular.

Hatchways closed by portable covers and secured weathertight by tarpaulins and battening devices.

a) the coamings of hatchways closed as above shall be of substantial construction and the height above the deck shall be at least as follows:

600 mm if in position 1
450 mm if in position 2

b) the width of each bearing surface for hatchway covers shall be at least 65 mm.

c) where covers are made of wood, the finished thickness shall be at least 60 mm in association with a span of not more than 1·5 m.

d) where covers are made of mild steel the strength has to be calculated with assumed loads the details of which are given in the Regulations.

e) Cleats shall be set to fit the taper of the wedges. They must be at least 65 mm wide and spaced not more than 600 mm centre to centre.

f) Battens and wedges to be efficient and in good condition. Wedges to be of tough wood and have a taper of not more than 1 in 6 and not less than 13 mm thick at the toes.

g) At least two layers of tarpaulin in good condition to be provided for each hatchway in position 1 or 2.

h) For all hatchways in position 1 or 2 steel bars to be provided to efficiently secure each section of hatchway covers after the tarpaulins are battened down. Figure 5–25(a) and (b).

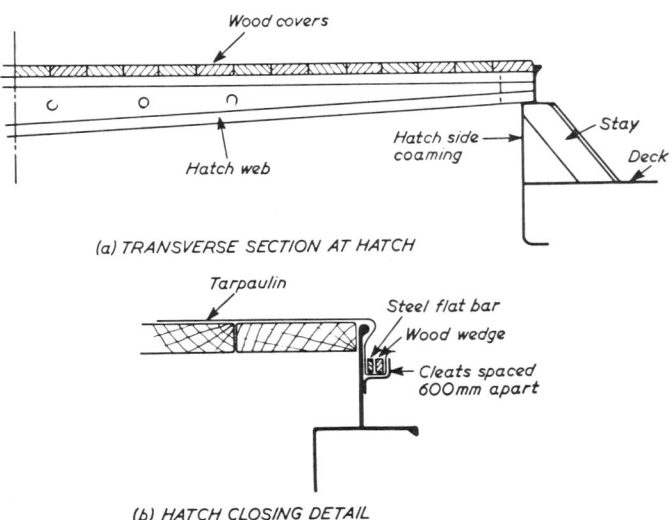

FIG. 5–25—*(a) Transverse section of hatch.*
(b) Hatch closing detail.

THE SHIP GIRDER AND STRUCTURAL DETAILS 69

Hatchways closed by weathertight covers of steel or other equivalent material fitted with gaskets and clamping devices.
a) At positions 1 and 2 the height above the deck of hatchway coamings to be as specified above. The height of the coamings may be reduced or omitted entirely on condition that the Administration is satisfied that the safety of the ship is not thereby impaired in any sea conditions. Where coamings are provided they shall be of substantial construction.

b) Where weathertight covers are of mild steel the strength has to be calculated with assumed loads, the details of which are given in the Regulations.

c) The means of securing and maintaining weathertightness shall ensure that the tightness can be maintained in any sea conditions and for this purpose tests are required at least at the initial survey.

A large number of present day ships are fitted with steel hatch covers. There are several types available but the self-supporting type is probably the most popular. The covers are arranged in four to six sections which extend across the hatchway and have rollers which operate on a runway. The covers are opened by rolling them to the end of the hatch where they tip automatically into the vertical position. Figure 5–26 shows a single pull fixed chain hatch cover designed by the MacGregor International Organization. All operations necessary for opening and closing the hatch covers are push button controlled.

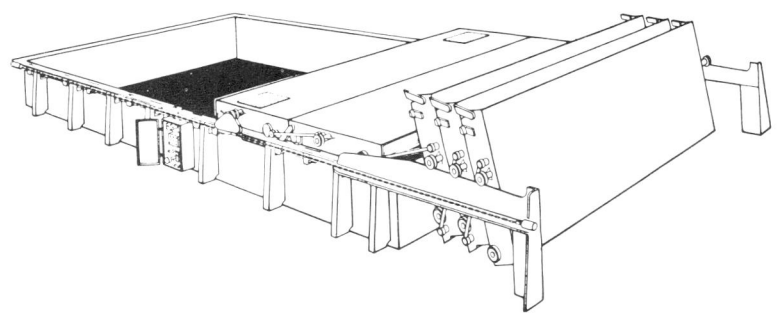

FIG. 5–26—*MacGregor single pull fixed chain hatch cover.*

Beams, Girders and Pillars
Where the deck is supported by transverse beams in association with longitudinal girders the beams are carried across the ship and connected to the side frames by beam knees. Figure 5–27. A continuous longitudinal girder is fitted on each side of the ship alongside the hatches. The beams are connected to the girders thus reducing their span.

Where deck-longitudinals are used they are supported by transverses which are bracketed to the frames. Figure 5–28.

In the cargo holds and 'tween decks support is given to the girders by pillars at the hatch covers. Tubular pillars are most often used in cargo spaces since they reduce cargo damage. In insulated holds the pillars are generally square. In ships with more than one deck the upper and lower pillars are arranged, as far as possible, in

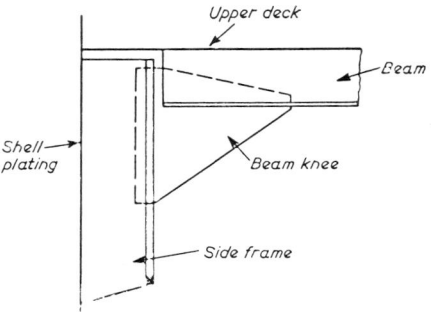

Fig. 5–27—*Beam knee.*

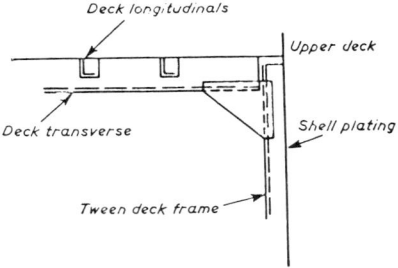

Fig. 5–28—*Deck transverse.*

the same vertical line, so that the downward pressure is direct. With end-on loads the strength of a pillar diminishes rapidly as the length is increased. Thus short pillars may be relatively small, whereas long pillars must be large to support even moderate loads.

Superstructures

A superstructure is considered as any structure built above the main continuous structure of the ship. These structures being less than the length of the ship introduce discontinuities in the ship girder. Typical superstructures are the poop, bridge, forecastle and deckhouses. With the poop, bridge and forecastle the sides are a continuation of the side plating of the main hull. With the deckhouse the sides are, in general, set in from the ship's side. A superstructure over a considerable part of the length amidships is of considerable value as far as longitudinal strength is concerned. A short bridge cannot be regarded as of much account as a strength member. Being attached to the side plating some of the hull stresses are transferred to its structure. The stresses are highest at the ends of the superstructure and the side plating should be gradually run into the hull so as to avoid structural discontinuity. With a long bridge at about 50 per cent of the length of the ship it can be assumed that the effective depth of the ship girder extends to the bridge deck. It is, however, important to appreciate that it is at the points where the strength deck steps up or down that the stresses are highest and local reinforcement of stringer and sheer strakes is necessary.

In large passenger ships with extensive superstructures the problem of mass becomes very important and it is in this range that aluminium alloys are being increasingly used. The mass of aluminium is about one-third that of steel and although the scantlings may require to be increased due to the lower tensile strength the saving in mass is appreciable and this in turn assists initial stability. Other features about the use of aluminium are low maintenance costs due to the metal's high resistance to a marine atmosphere and it is less dependent on protection by paint. Corrosion of aluminium does take place through the close proximity of certain other metals which leads to galvanic attack. Great care must therefore be taken when joining an aluminium alloy deckhouse to a steel deck.

Mild Steel and Higher Tensile Steel

Mild Steel

From near the end of the nineteenth century until the beginning of World War II, the conventional material for merchant shipbuilding was mild steel, and it was on this material that the rules of the classification societies were based.

Mild steel has an ultimate strength of about 46,500 N/cm²
modulus of elasticity about 20·9 MN/cm²
yield strength about 23,200 N/cm²

The advantages of mild steel are:
1) relative cheapness of production
2) high mechanical properties
3) ease of cold working
4) ability to be hot worked without loss of mechanical properties.

A considerable disadvantage is that it requires efficient protection to prevent corrosion.

With the introduction of welding in lieu of riveting, it eventually became possible to weld mild steel comparatively easily.

Brittle Fracture

In certain welded ships in the 1940s operating at low climatic temperatures, a phenomenon known as brittle fracture occurred; the steel fractured at nominal stresses which were much below what would be considered to be the working stress for the material.

Such an occurrence brought about intensive investigations of the problem in addition to detailed study of the actual stress experienced by ships and the effect of the composition of the material and its heat treatment on the properties of mild steel.

The problem was more severe in welded ships than in those which had been riveted. This was due to the continuity of the structure in the welded ship where there was no barrier to stop a crack; in a riveted ship a seam would act as an arrester since the crack would have to start in the next strake of plating before it could continue. Brittle fracture in mild steel depends largely on temperature. There is a certain critical temperature above which it does not take place.

A steel is now produced that will not suffer brittle fracture at the temperature at which it is used. Such steels are known as notch tough steels. Classification societies now require the main structural strength members of ships which will be subjected to high stress in service to be of steel which has high resistance to brittle fracture. The unified requirements specify five grades of steel designated as A, B, C, D and E. The ultimate tensile strength is identical for all grades. Charpy V notch impact tests are required for grades D and E. All these steels meet the low temperature problems arising from climatic conditions. Further details are given in a paper by Boyd and Bushell, Transactions RINA, 1961.

Steels for very low temperatures are referred to in Chapter 4.

Higher Tensile Steel

The term "higher tensile steel" is used in shipbuilding to avoid confusing shipbuilding steels with the really high tensile steels which can have ultimate strengths of 150,000 N/cm^2. The higher strength does not affect the modulus of elasticity and the density, these being, as for mild steel, in the region of 20·9 MN/cm^2 and 7,850 Kg/m^3 respectively. The ultimate strength on average is about 55,000 N/cm^2 so that the increase in strength is of the order of 20 per cent compared with mild steel. Even at that level, the saving in mass can be quite extensive.

The main use of such steel is in the large ships such as oil tankers, ore carriers, bulk carriers, and the large passenger liner. In these very large ships, considerable thickness of material would be required in the main longitudinal strength members if the ordinary mild steel were used; thus by the adoption of higher tensile steel, the thickness of material can be reduced.

THE SHIP GIRDER AND STRUCTURAL DETAILS 73

For ship structures, steel, in addition to strength, must be weldable and have a good notch toughness.

Aluminium Alloys

Pure aluminium has a good resistance to corrosion due to the formation of a protective film of oxide. However, the pure metal is too soft for constructional purposes. There are, however, a number of special alloys, some of which have tensile strengths equal to that of steel. In general, very high-strength alloys are not suitable for shipbuilding; the alloys actually suitable for shipbuilding are few in number.

Aluminium alloys are, in general, classified into two types: heat treatable and non-heat treatable, and both types are accepted by the classification societies for use in ship's structures.

An important property of aluminium alloy is the modulus of elasticity which is more or less constant for all such alloys. The value is $6 \cdot 98$ MN/cm^2, and is about one-third the value for steel. An interesting point arises when the structural strength members of a vessel embody materials of very different properties. Suppose steel with a modulus of $20 \cdot 9$ MN/cm^2 is associated with aluminium of $6 \cdot 98$ MN/cm^2 then equality of stress is obtained by calculating the section modulus on a basis of one-third of the actual area of aluminium, i.e.: $E/E_1 \approx 7/21 = \frac{1}{3}$.

If steel is replaced by an aluminium alloy, then the net saving is generally in the region of 50 per cent of the steel mass. The mass saving advantage of aluminium over mild steel shows itself in the superstructure of passenger ships. Moreover, the low modulus of elasticity of aluminium lends itself to relieving the stresses in long superstructures.

Aluminium alloys are readily available in both plate form and bar extrusions. The extrusion process enables a great variety of shapes of sectional material to be produced. Normal shipyard machinery and working methods can be used for this material.

The film of aluminium oxide which gives the metal its high corrosive resistance did create early welding difficulties; these, however, were overcome by the development of the inert-gas shielded arc process which has become the established method for large ship structures. Due to high heat penetration, good quality welds are possible.

Aluminium alloys are below steel in the electro-chemical series, thus they are electro-negative to steel and corrode first. The potential difference between aluminium alloys and steel is quite large, and thus special precautions have to be taken where the two metals come in contact. Tape impregnated or heavily coated with zinc oxide paste is a suitable separating medium.

Corrosion

The investigation of methods to prevent the deterioration of ships' hulls during service is a continuous process that has been going on for many years. As far back as 1824 Sir Humphrey Davy reported that cathodic protection could be applied to copper sheathing on timber ships by the use of iron anodes. However, the overall protection of seagoing ships was not explored in detail until the early 1950s when it was shown that cathodic protection in conjunction with suitable coating systems could make considerable savings in maintenance costs.

The cost of corrosion and prevention within the marine industries is estimated at £280 million a year. In view of this the advisability of protecting vessels against corrosion is not questioned and investigations continue into the merits of various coating materials and the techniques available.

The problems involved in the corrosion protection of steel plate are, in general, similar irrespective of type of ship although in the case of tankers the corrosion problem exists not only on exterior steel plates but also in cargo spaces. There are considerable economic advantages to be derived from using modern methods of steel preparation coupled with an efficient paint system.

Corrosion and consequent roughening of the hull leads to reduced speed and/or increased fuel consumption. This loss has been estimated at between 6 and 10 per cent within two years. Some years ago a table was published showing that over a period of four years service a tanker for the same speed required more power and had a substantial increase in the daily fuel consumption. This was not due to variation in weather, or propeller. The resistance of the ship had changed and had caused an increase in friction between ship and water. An investigation at the time indicated that the fall in performance was largely due to corrosion.

Experience has shown that superior long-term performance of a ship can be achieved by careful corrosion protection, provided consideration is given to the factors under which superior performance can be expected.

So long as ships are built of steel they will corrode, but corrosion control systems for steel have made very great progress in the last few years. The whole basis of the developments in protection against corrosion has been the improved methods of preparing the steel surfaces for the coating materials. There is no doubt that the basic factor is the condition of the steel to be treated. One of the most important types of corrosion is caused by mill-scale, the bluish-black scale of iron oxide which forms on the plates when they are being hot-rolled at the steel works. It has long been recognized that the presence of mill-scale leads to local pitting. In earlier methods of ship construction the plates were stored at the shipyards to weather in order to loosen some of the scale. During the subsequent working of the plate the breakdown of the scale continued. The plating was then wire brushed and painted. This laborious operation invariably left a great deal of the mill-scale intact in addition to a pitted surface. Accelerated removal of the mill-scale can be achieved chemically or mechanically. With the chemical method the scale and rust are removed by means of mineral acids, but for large objects such as ship's plates there are problems associated with the installation and maintenance of large acid baths.

The standard method now adopted for removing mill-scale and rust is to shot blast the plate and immediately apply a shop priming coat. Various types of abrasive can be used, such as sand, grit, slags, shot and chopped steel wire. Cleaning steel in this way, prior to cutting and welding, makes it possible to observe faults in the steel at an early stage.

The shop priming coat, to perform effectively, must comply with a number of requirements such as:
1) Be dry to handle within a few minutes
2) Protection against corrosion during the period of building
3) Able to resist damage in the event of rough handling
4) Resistant to the high temperatures associated with cutting and welding
5) Safe level of toxic vapours during cutting and welding
6) Capable of overcoating with the normally used marine paint.

The main purpose of the shop primer is the protection of steel during construction. Full corrosion protection, however, can only be achieved by overcoating with an adequate film thickness of suitable top coats.

The oil tanker presents a serious corrosion problem since not only the external hull requires protection but also the cargo tanks which in some cases have to take corrosive liquids. The methods of tank protection are cathodic protection, water soluble inhibitors, oil soluble inhibitors and coatings, It has been claimed that epoxy coated tanks, although of higher first cost, show advantages over the other methods.

Lloyd's permit scantling reductions in certain items of structure provided an approved system of corrosion control is adopted. In one type of 100,000 tonne deadweight tanker the saving in steel was estimated to be about 600 tonnes.

Lloyd's rules require that all practicable steps should be taken to remove all mill-scale before coating the external steelwork and it is recommended that shot blasting or some other equally effective method should be employed. It is also required that in general all steelwork should receive at least two coats of paint, except inside tanks intended for oil.

Fouling

In addition to the prevention of corrosion it is essential to prevent or at least reduce as far as possible the fouling of the under water plating by marine animal and vegetable growths; these increase the resistance of the ship, reduce speed and increase fuel consumption. Cathodic protection does not prevent fouling. Special anti-fouling compositions containing poisons are used for this purpose and one or more coats are usually applied over the protective composition.

No matter how carefully bottom paint has been applied, experience has shown that new ships should be drydocked for cleaning and recoating within six months after launching.

Lloyd's consider it desirable to drydock at about 12-monthly intervals. The maximum interval is two years. Ordinary anti-fouling compositions have an effective life of about 12 months after which they lose the poisonous effect and permit marine growth and the underlying paint may begin to peel off and start corrosion.

Cathodic Protection

If two exactly similar metals in metallic contact are placed in pure distilled water there will be no passage of electricity between them even if they be linked with an external supply of current. Immediately a small quantity of impurity is added to the water a flow will commence. The liquid has now become an electrolyte and the process so established is known as "electrolysis" and electrolytic corrosion takes place.

Electromechanical corrosion occurs when two dissimilar metals are present in an electrolytic medium. Sea water is a most efficient electrolyte. Different parts of the same metal made dissimilar such as by work done on the metal or a metal and its oxide are in themselves sufficiently dissimilar as to create corrosion. An anodic area, such as iron oxide, is eaten away creating more rust, while an electric current is created, leaving the metal at the anodic area and entering at the cathodic area, where no corrosion occurs.

An efficient form of cathodic protection is offered by opposing the natural effect by impressing a direct current in the circuit and using anodes which may or may not corrode. The entire immersed surface thus becomes cathodic. This passive system is called impressed current cathodic protection and is suitable for large areas such as liquid cargo tanks and the outside of the hull.

CHAPTER 6

Statutory Regulations

Statutory Regulations; *Classification Societies*; *Freeboard*; *Tonnage*; *Registration*

Statutory Regulations

Almost all merchant ships are built under the rules of a classification society. All must comply with the statutory regulations. Statutory regulations are laid down by the Department of Trade and Industry (DTI) as the British Government's authority for declaring the standards of safety for merchant ships, related to damage, collision, subdivision, life-saving equipment, loading, stability, fire protection, navigation, carriage of dangerous cargoes, load lines and other allied subjects. The DTI is also the authority on tonnage measurement as well as ensuring that safety measures, many of which are controlled by international agreements, are maintained.

Thus, the primary object of statutory regulations is to promote safety of life at sea. The rules issued by the DTI are, as stated, compulsory and are enforced by the various Merchant Shipping Acts. It is the purpose of the Government to ensure that the standards appropriate to safety are adopted. The international scope of the operations of ships has considerable bearing on the interrelation between the regulations of different governments. Comparable uniformity on an international scale has been made by means of Conferences, at which conventions were formulated. This is clearly indicated by the International Load Line Convention, the International Convention for the Safety of Life at Sea and the International Conference on Tonnage Measurement of Ships.

Some idea of the function of the DTI may be obtained from the following list of rules issued by the Department. The rules are intended to ensure the safety of ships and of those who travel in them.

1) *Load Line Rules*

These are the rules agreed to under the International Convention. Freeboards are assigned according to the geometrical properties of the ship and the structural strength, in conjunction with the strength and security of covers to the deck openings and other considerations.

2) *Survey of Passenger Ships*

A ship intended to carry more than 12 passengers must conform to the regulations for passenger ships, and be issued with a Passenger Certificate. All passenger ships must be surveyed annually for a renewal of this certificate and the survey requires the hull to be examined in dry-dock. The Passenger Certificate must be posted up in a prominent and accessible place in the ship.

Every passenger ship and cargo ship must be inclined upon completion, in the presence of a DTI surveyor, to determine the elements of her stability. From the results, the ship's master must be supplied with information for guidance in the loading and ballasting of the ship.

3) *Life Saving Appliances (LSA)*

In general, passenger ships are required to carry lifeboats under davits for all persons on board and buoyant apparatus for an additional percentage of the number on board. Cargo ships, generally, are required to be provided with lifeboats under davits on each side of the ship to accommodate all persons on board.

All lifeboats must be built to conform to the requirements of the LSA Rules and are inspected during construction.

All persons on board all ships must be provided with an approved lifejacket.

4) *Masters and Seamen (Crew Accommodation)*

The Merchant Shipping Acts lay down certain minimum standards for crew spaces regarding floor area, construction, lighting, heating, ventilation etc. Plans and details of accommodation must be submitted to the Department at an early stage for approval. The actual accommodation is finally inspected and measured at the ship.

5) *Tonnage Measurement*

Tonnage measurement is required to establish the gross and net tonnages upon which port and harbour dues etc. are levied.

6) *Grain Cargoes*

Regulations are laid down so as to limit the effect of any transverse shift of grain.

Certain trim and stability conditions must be approved.

7) *Lights and Sound Signals*

International regulations for preventing collision at sea require the provision of proper navigation lights and means of making sound signals in all ships. The lights must be screened at the ship so that they will show only in specified directions.

8) Fire Appliances

The provision of arrangements for the prevention, detection and extinguishing of fire on board ship are most extensive. The means adopted may be divided into three parts, namely (a) fire-proofing the ship as far as possible, (b) means for detecting a fire where and when it occurs and (c) means for extinguishing fires. In passenger ships fire patrols must be kept, and an alarm and detecting system fitted. Extinguishing arrangements by means of jets of water from hoses and portable fire-extinguishers are required.

Fire-smothering gas or smothering steam is provided in the cargo spaces by permanent piping systems.

Special arrangements are required in the propelling machinery spaces.

Classification Societies

There is little doubt that the classification societies have a profound influence on shipping, ship design and ship safety. The fundamental purpose of classification is to ensure maintenance of seaworthiness of all classed ships.

The principal maritime nations have the undernoted classification societies:

Great Britain	Lloyd's Register of Shipping
United States of America	American Bureau of Shipping
France	Bureau Veritas
Germany	Germanischer Lloyd
Norway	Det Norske Veritas
Italy	Registro Italiano
Russia	Register of Shipping of the USSR
Japan	Nippon Kaizi Kyokai
Poland	Polish Register of Shipping

The Classification Societies operate throughout the world and publish rules and regulations directly related to the structural efficiency of the ship and the reliability of the propelling machinery. Classification implies that a ship and the machinery conform to the standards published in the rules of the society. Classification is purely voluntary on the part of the shipowner, and the only penalty that can be imposed for non-compliance with the Rules is suspension or cancellation of class.

It is rather interesting that the two extremely influential bodies in the industries of shipbuilding and shipping should derive their names from the same owner of a coffee shop, Edward Lloyd. These two bodies, Lloyd's Register of Shipping and Lloyd's Insurance Corporation, are entirely independent. The latter, Lloyd's Insurance Corporation, is concerned with mercantile and other insurance business whereas Lloyd's Register of Shipping is concerned with the maintenance of standards in ship construction and the classification of ships. When a ship is submitted for insurance, underwriters require a guarantee that the vessel is structurally sound for the intended service. It was for the purpose of supplying the guarantee that several classification societies were established.

Lloyd's Register of Shipping was founded in 1760 and reconstituted in 1834. At the reconstitution a definite system of classification was established. Classification with Lloyd's entails approval of constructional plans, testing of materials, construction under survey, and recommendation for class from the surveyors by report to the Committee. Following the acceptance of the report by Lloyd's Committee, the certificate of class is issued and the record made in the Register Book.

The highest class given by Lloyd's is ✠ 100 A1. New ships are required to be built under Special Survey and the cross is inserted before the classification numeral in the Register Book. The figure 1 after the character of classification indicates that the equipment of anchors, cables and hawsers is in good condition. When the class 100 A1 is assigned it may be followed by the descriptive notation, oil tanker, ore carrier, etc. Maintenance of standard is an important function of Lloyd's Register. Periodical surveys are required and failure to conform may result in removal of the ship from class and a consequent reduction in its value.

Annual Surveys. All steel ships should be surveyed at intervals of approximately one year in accordance with the Rules. These annual surveys should, where practicable, be held concurrently with statutory annual or other load line surveys.

Docking Surveys. A ship should be examined in dry dock at intervals of about 12 months; the maximum interval is two years.

Special Surveys. All steel vessels classed with Lloyd's are to be subjected to Special Surveys in accordance with the Rules. These surveys become due at five-yearly intervals, the first five years from the date of build or date of Special Survey for Classification, and thereafter five years from the date of the Special Survey.

The date of completion of the Special Survey during construction of ships built under the Society's inspection is normally taken as the date of build.

Lloyd's Machinery Certificate (LMC)

Machinery constructed and installed on board in accordance with the rules and on satisfactory completion of trials is assigned the class notation LMC (Lloyd's Machinery Certificate). New machinery for ships intended for classification is to be constructed under the Society's

Special Survey and on completion will have the mark ✠ inserted before the machinery class notation. Thus ✠ LMC.

The standards to which the ships must be built and maintained are laid down in the Lloyd's Register publication "Rules and Regulations for the Construction and Classification of Ships". This is issued annually and constantly being revised to meet new demands. The Lloyd's Register, issued in two volumes, is a supreme work of reference, used and trusted throughout the world. It contains a list as complete as can be ascertained of all sea-going merchant ships in the world of 100 tons gross and above whether classed by Lloyd's Register or not. The number of ships thus listed exceeds 50,000.

The Society is empowered to assign load line certificates to ships and to ensure that the conditions related to them are carried out. It is, thus, apparent that a classification society, such as Lloyd's Register of Shipping, is not only able to assist to an enormous extent in making ships safe to travel the seas, but is able to accumulate a vast amount of information on the behaviour of the structures under sea-going conditions; this information serves to suggest means whereby the structures of ships can be improved. Lloyd's Register of Shipping carries out a large amount of research work, which also suggests lines of improvement in the structure of ships.

Freeboard

In general terms freeboard is the amount of the side of the ship out of the water. The minimum freeboard is the height amidships of the freeboard deck at side above the normal summer load-line. The fixing of the position of the load line, and the measurement of tonnage, constitute two very important factors. The former controls the paying load and the latter assesses a quantity which is involved in the cost of running a ship.

The fundamental reason for statutory enforcement of a limiting load line was to protect seamen against the risk of proceeding to sea in unseaworthy ships. The name of Samuel Plimsoll will always be associated with the subject of freeboard; he made a successful fight in Parliament for the marking of all ships with a line indicating the limit to which they could be loaded. A number of Acts were passed and eventually, in 1876, a Bill was introduced whereby the owner was required to have the ship marked with a deck line and a circular disc on the ship sides indicating the maximum draught to which it was proposed to load. This disc was popularly known as the Plimsoll mark. The Bill, however, did not include rules whereby the freeboard could be estimated; the position of the disc was left to the discretion of the owner.

Several suggestions about freeboard were made over the years, all without legal obligation, until the Merchant Shipping Act of 1890, which required all British ships, excluding fishing vessels, yachts etc., to have freeboards assigned in accordance with tables and regulations. This operated until 1906 when a major revision was made, and this continued until the Merchant Shipping Act of 1932. This Act was the result of the International Convention of 1930 and for the first time in the history of the world's shipping common agreement was reached on a uniform method of applying freeboards to merchant ships.

1966 Load Line Convention

The current rules governing the allocation of freeboard are laid down by the International Load Line Convention of 1966 and ratified by each of the countries taking part and by the United Kingdom in the Merchant Shipping (Loadline) Rules 1968. The Conference at which the Convention was signed was held in London during 1966 upon the invitation of the Inter-Governmental Maritime Consultative Organization (IMCO). Further details of IMCO and its activities are given in Chapter 7.

The statutory calculation of freeboard involves water density, the ship's length, breadth, depth, sheer, extent of watertight superstructures and other geometrical features of the ship. Standards are laid down for hatch covers, crew protection, freeing port areas, ventilators etc.

The load line rules assume that the nature and stowage of the cargo, ballast etc., are such as to secure sufficient stability of the ship and the avoidance of excessive structural stress.

Ships built and maintained in conformity with the requirements of a recognized classification society may be considered to possess adequate strength.

Freeboard Deck is normally the uppermost deck exposed to weather and sea, which has permanent means of closing all openings in the weather part and below which all openings in the sides of the ship are fitted with permanent means of watertight closing.

Deck Line is a horizontal line 300 mm in length and 25 mm in breadth. It is marked amidships on each side of the ship and its upper edge passes through the point where the upper surface of the freeboard deck intersects the outer surface of the shell as shown in Figure 6–1.

Load Line Mark consists of a ring 300 mm in outside diameter and 25 mm wide intersected by a horizontal line 450 mm in length and 25 mm in breadth, the upper edge of which passes through the centre of the ring. The centre of the ring is placed amidships and at a distance equal to the assigned summer freeboard below the upper edge of the deck line as shown in Figure 6–2.

The undernoted load lines are used:

Summer load line passes through the centre of the ring and marked *S*.
Winter load line indicated by upper edge of line marked *W* and is
    obtained by adding to the summer freeboard one forty-eighth of
    the summer draught.

STATUTORY REGULATIONS

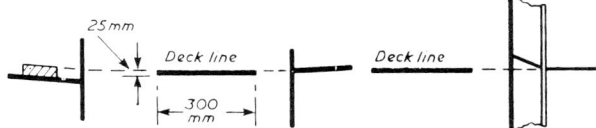

Fig. 6–1—*Deck line.*

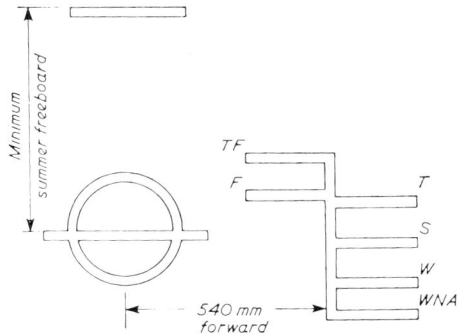

Fig. 6–2—*Load line mark.*

Winter North Atlantic load line indicated by upper edge of line marked *WNA*. For ships not more than 100 metres in length it is the winter freeboard plus 50 mm. For other ships the *WNA* freeboard is the winter freeboard.

Tropical load line indicated by upper edge of line marked *T* and is the summer freeboard less one forty-eighth of the summer draught.

Fresh Water load line indicated by upper edge of line marked *F* and is the summer freeboard less $\Delta/40T$ cm where Δ = displacement in sea water in tonnes at the summer load line and T = tonnes per centimetre immersion in sea water at the summer load water line.

Tropical Fresh Water load line indicated by the upper edge of line marked *TF* and is the fresh water freeboard less one forty-eighth of the summer draught.

There are special markings for timber freeboards and sailing ships.

The initials of the authority by whom the load lines are assigned are indicated alongside the load line ring. The ring, lines and letters are painted in white or yellow on a dark ground or in black on a light ground. They are also permanently marked on the sides of the ships.

For the purposes of freeboard computation ships are divided into Type "A" and Type "B".

Type "A" is the ship designed to carry only liquid cargoes in bulk and if over 150 metres in length must have a certain standard of subdivision; if over 225 metres in length the machinery space must be taken into account for subdivision computation.

Type "B". All ships which do not come within the provisions regarding Type "A" ships are considered as Type "B".

The actual computation for the assignment of freeboard is a relatively simple calculation and consists of corrections made to a basic freeboard. This basic freeboard is dependent on the length and type of vessel and has considerable influence on the final calculated value. The calculation is generally carried out, like many calculations in naval architecture, in tabular form on a standard sheet. The worked example, given later, is shown on such a tabular form.

Some of the terms used are defined thus:

Length (L)

The length is taken as 96 per cent of the total length on a waterline at 85 per cent of the least moulded depth *or* as the length from the fore side of the stem to the axis of the rudder stock on that waterline, if that be greater.

Breadth (B)

The maximum breadth of the ship, measured amidships, to the moulded line.

Depth Moulded

This is the vertical distance measured from the top of the keel to the top of the freeboard deck beam at side.

Displacement

The moulded displacement of the ship, excluding bossing, taken at a moulded draught d_1 which is 85 per cent of the least moulded depth.

Block Coefficient (C_b)

is derived from the above displacement in association with L, B and d_1.

Freeboard Deck is normally the uppermost complete deck exposed to weather and sea which has permanent means of closing all openings in the weather part and below which all openings in the sides of the ship are fitted with permanent means of watertight closing.

STATUTORY REGULATIONS 85

Superstructures. These are the detached superstructures *on* the freeboard deck, of efficient construction, and extending transversely to at least within 0·04 B from the ship's sides. The length of a superstructure (S) is the mean length of the part of the superstructure which lies within the length (L). Details of the length (S) and the effective length (E) are given in Regulations 34 and 35. The standard heights of superstructures, in metres, are as under:

L (metres)	RQD	All other Superstructures
30 or less	0·90	1·80
75	1·20	1·80
125 or more	1·80	2·30

The actual assessment of freeboard, as stated above, is made up of a series of corrections to a basic freeboard. These basics for the types "A" and "B" are given in Regulation 28 of the Convention. The corrections are as follows:
 1) For ships of type "B" between 24 metres and 100 metres in length.
 2) Block coefficient.
 3) Depth.
 4) Extent of superstructures.
 5) Sheer.
A summary of these is now given.

1) *Correction to Type "B"*
The tabular freeboard for a type "B" ship of between 24 metres and 100 metres in length having enclosed superstructures with an effective length of up to 35 per cent of the freeboard length (L) is increased by

$$7·5 (100 - L) (0·35 - E/L) \text{ mm}$$

2) *Block Coefficient (C_b) Correction*
Where C_b exceeds 0·68 the tabular freeboard—as modified, where necessary, by item (1) above—is multiplied by

$$\frac{C_b + 0·68}{1·36}$$

3) *Depth Correction*
The freeboard depth (D) is the depth moulded amidships plus thickness of freeboard deck stringer plate plus

$$\frac{T(L - S)}{L}$$

as per Regulation 3 (6).

Where D exceeds $L/15$ the freeboard is *increased* by $(D - L/15)$ R mm where R is $L/0 \cdot 48$ at lengths less than 120 metres and 250 at 120 metres length and above. Where D is less than $L/15$ no reduction is made except where an enclosed superstructure covers at least $0 \cdot 6 L$ amidships as per Regulation 31 (2). Where the height of superstructure or trunk is less than standard height the reduction is in the ratio of the actual to the standard height—Regulation 31 (3).

4) *Superstructure Correction*

Where the effective length of superstructures and trunks is $1 \cdot 0 L$ the freeboard can be *reduced* by 350 mm at 24 metres length of ship, 860 mm at 85 metres length and 1,070 mm at 122 metres length and above, deductions at intermediate lengths to be obtained by linear interpolation. Where the total effective length is less than $1 \cdot 0 L$ the deduction is a percentage as given in Tables of Regulation 37 (2). For ships of type "B" special conditions apply as per Regulation 37 (3).

5) *Sheer Correction*

The standard parabolic sheer curve has an aft ordinate (S_a) given by $25 (L/3 + 10)$ mm and the forward ordinate (S_f) is $2S_a$. The area under the actual curve is compared with that under the standard curve. Where the actual or virtual sheer profile differs from the standard, details are given in Regulations (10) and (11).

A summary is given below.

Actual or Virtual Sheer as compared with		Allowable Sheer	
Aft	Forward	Aft	Forward
Deficient	Deficient	Actual	Actual
> 0·75 Standard	Excess	Actual	Actual
< 0·50 Standard	Excess	Actual	Standard
> 0·50 < 0·75 Standard	Excess	Actual	Stnd. + Inter. Allow.
Excess	Excess	Actual	Actual
Excess	Deficient	Standard	Actual

The correction for deficiency or excess of sheer is the difference between the actual sheer and the standard sheer multiplied by

$$\left(0 \cdot 75 - \frac{S}{2L}\right)$$

where S is the total length of enclosed superstructure. Where the sheer is less than standard the correction is *added* to the freeboard. If the

STATUTORY REGULATIONS 87

sheer is in excess of standard a *deduction* may be made from the freeboard if superstructure covers $0{\cdot}1\,L$ abaft and $0{\cdot}1\,L$ forward of amidships. The deduction for excess of sheer may be made if no superstructure covers amidships. Where superstructure covers less than $0{\cdot}1\,L$ abaft and forward of amidships the deduction is obtained by linear interpolation. The maximum deduction for excess sheer is at the rate of 125 mm per 100 metres length.

When the foregoing corrections are made to the basic freeboard the resulting freeboard will correspond to the maximum geometric summer draught for the type of ship considered. The freeboard thus calculated may, however, be increased by the undernoted if:
  a) the bow height is insufficient—Regulation 39;
  b) the structural strength of the hull is not sufficient for the draught—Regulation 1;
  c) the sub-division is unsuitable—Regulation 27;
  d) cargo ports or other similar openings are arranged in the sides of the ship below the freeboard deck—Regulation 21;
  e) the owners desire an assignment of freeboard corresponding to a draught less than the maximum geometric draught.

The special requirements for ships assigned timber freeboards are given in Chapter IV—Regulations 41 to 45—of the Convention.

EXAMPLE

A vessel has been selected of Type "B" to indicate the procedure in the computation of freeboard.
The principal particulars are as follows:
Length on waterline at 85 per cent of least moulded depth $= 246$ m;
Length on waterline at 85 per cent of least moulded depth to centre line of rudder stock $= 240$ m;
Breadth moulded $(B) = 32$ m; Depth moulded amidships $= 19$ m;
Least moulded depth $= 19$ m;
Thickness of stringer plate $= 35$ mm; thickness of keel $= 30$ mm;
No deck sheathing;
Displacement moulded at moulded draught 85 per cent of least depth moulded $= 104{,}200$ tonnes;
Length of Forecastle $= 17{\cdot}5$ m; Height $= 2{\cdot}5$ m;
Length of Bridge $= 15{\cdot}0$ m; Curved $1{\cdot}0$ m; Height $= 2{\cdot}5$ m; Width $(b) = 30$ m;
Length of Poop $= 21{\cdot}9$ m; Curved $1{\cdot}0$ m; Height $= 2{\cdot}5$ m;
Poop height at AP $= 3{,}000$ mm; Forecastle height at FP $= 3{,}000$ mm;
Sheer ordinates at intervals of one-sixth of freeboard length from the after terminal: 400, 20, 0, 0, 0, 1,700 mm.

The calculation is given in Table 6.1 and subsidiary items are indicated below:

$$\text{Table depth} = L/15 = \frac{240}{15} = 16 \text{ m}$$

$$\text{Block coefficient} = \frac{V}{LBd_1} = \frac{104{,}200 \times 0{\cdot}975}{240 \times 32 \times 19 \times 0{\cdot}85} = 0{\cdot}82$$

TABLE 6.1.
COMPUTATION OF FREEBOARD

Ship's Name: ; Builder's No.:   ; Date:

Length on waterline (L_W) at 85 per cent of least moulded depth $\begin{cases} L_W \times 0{\cdot}96 = 246 \times 0{\cdot}96 = 236{\cdot}1 \text{ m} \\ \text{To centre line of rudder stock} = 240 \text{ m} \end{cases}$

Freeboard length (L) = 240 m; Breadth moulded (B) = 32·0 m; Depth moulded $\not{D}$ = 19·0 m

Disp. moulded at 85 per cent depth moulded = 104,200 tonnes; Block coefficient = 0·82; Least moulded depth = 19·0 m

DEPTH FOR FREEBOARD (D)	DEPTH CORRECTION
Depth moulded = 19·000	a) Where D is greater than table depth [$L/15$]
Stringer plate = 0·035	$[D - L/15] R = [19{\cdot}035 - 16{\cdot}000]\ 250 = 759$ mm
Sheathing on exposed deck = — = —	b) Where D is less than $L/15$ (if allowed)
$7\left[\dfrac{L-S}{L}\right]$ =	If restricted by superstructure
Depth for freeboard [D] = $\overline{19{\cdot}035}$	

DEDUCTION FOR SUPERSTRUCTURES

Superstructure	Mean covered length (S)	Height	Height correction	Effective length (E)	
Forecastle	17·50	2·5	—	17·50	Standard height of superstructures and trunks = 2·30 m
Bridge	15·66	2·5	—	14·68	Standard height of RQD = —
Poop	22·56	2·5	—	22·56	Deduction for complete superstructures = 1070 mm
RQD	—	—	—	—	Percentage covered S/L = 23·22
Trunk Aft	—	—	—	—	Percentage covered E/L = 22·81
Trunk Forward	—	—	—	—	
Total	55·72			54·74	

TYPE "A" SHIP

Per cent from Table = —   ; Deduction = —

TYPE "B" SHIP

	Line I	Line II
Percentage from Table	11·41	14·47
Corrected for forecastle less than 0·07 L	—	—
Interpolation for bridge less than 0·2 L	—	12·41
Deduction	$1070 \times 0·1241 = 133$ mm	

Forecastle = 0·073 L

Bridge = 0·061 L

SHEER CORRECTION

Sum of standard sheer aft products $= S_A \times 8/3 = 6{,}000$

Sum of standard sheer forward products $= S_F \times 8/3 = 12{,}000$

Ratio $\dfrac{\text{virtual sheer}}{\text{standard sheer}}$ $\begin{cases}\text{Aft} = \text{Deficient}\\ \text{Forward} =\end{cases}$

Length of enclosed superstructure $\dfrac{}{L}$ $\begin{cases}\text{Forward } \cancel{\boxtimes} =\text{ not}\\ \text{Aft } \cancel{\boxtimes} = \text{ applicable}\end{cases}$

Excess height of end superstructures:

Poop or RQD at AP $= 3{,}000 - 2{,}300 = 700$

Forecastle at FP $= 3{,}000 - 2{,}300 = 700$

$S = Y/3\ L_1/L$	
$700/3 \times 17·5\ /240 = 17·01$	S_f
$700/3 \times 22·56/240 = 21·93$	S_a

STATUTORY REGULATIONS

Station	Actual ordinate	SM	Products
AP	400	1	400
1/6 L from AP	20	3	60
2/6 L from AP	0	3	0
Amidships	0	1	0

Sum = 460
Add $(16 \times S_f) = 351$

Sum of virtual sheer = 811

Allowable sum (a) = 811

Station	Actual ordinate	SM	Products
Amidships	0	1	0
2/6 L from FP	0	3	0
1/6 L from FP	0	3	0
FP	1,700	1	1,700

Sum = 1,700
Add $(16 \times S_f) = 272$

Sum of virtual sheer = 1,972

Allowable sum (b) = 1,972
Allowable sum (a) = 811

Total = 2,783

$$\text{Correction} = \frac{\text{difference between sums of products}}{16}\left[0{\cdot}75 - \frac{S}{2L}\right]$$

$$= \frac{12{,}000 + 6{,}000 - 2{,}783}{16}[0{\cdot}75 - 0{\cdot}1161] = 603 \text{ mm}$$

If limited on account of midship superstructure   No.

If limited to maximum allowance of 125 mm per 100 m of length: No.

BOW HEIGHT

Minimum bow height:

For ships below 250 m length $= 56 L\left[1 - \dfrac{L}{500}\right]\dfrac{1{\cdot}36}{C_b + 0{\cdot}68}$ mm $= 56 \times 240 \left[1 - \dfrac{240}{500}\right] \times \dfrac{1{\cdot}36}{0{\cdot}82 + 0{\cdot}08} = 6{,}337$ mm.

For ships 250 m length and above $= 7\,000 \times \dfrac{1{\cdot}36}{C_b + 0{\cdot}68} =$ —

Actual bow height $=$ depth $[D]$ + sheer at FP + forecastle height at FP $-$ draught moulded $=$

TABULAR FREEBOARD AND SUMMARY OF CORRECTIONS

		+	−	
Tabular freeboard	mm			3,880
Correction for superstructure length $< 0.35 L$				—
Corrected for $C_b = \dfrac{0.82 + 0.68}{1.36}$	=			4,279
Depth correction		759	—	
Deduction for superstructure		—	133	—
Sheer correction		603	—	
Thickness of deck		—	—	—
Position of deck line		—	—	—
Scantlings; Floodability		—	—	—
Bow height		1,362	133	+1,229

Summer freeboard = 5,508 mm
            = 5·508 m

		m
Depth to freeboard deck	=	19·035
Summer freeboard	=	5·508
Draught moulded (d)	=	13·527
Keel thickness	=	0·03
Extreme draught	=	13·557

Forecastle:
   mean covered length $(S) = 17·5$ m; $0·07 L = 16·8$ m; $0·04 L = 9·6$ m
   effective length $(E) = 17·5$ m
Bridge:
   mean covered length $(S) = 15 + 2/3 \times 1 = 15·66$ m; $0·2 L = 48$ m
   effective length $= 15·66 \times 30/32 = 14·68$ m
Poop:
   mean covered length $(S) = 21·9 + 2/3 \times 1 = 22·56$ m
   effective length $= 22·56$ m

The International Convention on Load Lines, 1966, will, because of the general increases in draught, increase the carrying capacity of the merchant fleets of the world; but due to the greater protection afforded by steel weathertight hatches, the new sub-division requirements and the special requirements for small ships, it will also improve their safety.

Tonnage

Tonnage is a very important subject associated with the shipping industry since it determines a quantity which is involved in the cost of running a ship. Tonnage measurement is required to establish the gross and net tonnages upon which port and harbour dues, etc., are levied.

The principle upon which British tonnage measurement is based has been the subject of controversy for generations. The earliest record in Britain relating to tonnage is in the year 1422 when a Government law stipulated that coal carrying vessels had to be measured and marked, but no reference was made as to how this was to be carried out. The term "tonnage" appears to have come into use when the dues then charged were based on the number of casks of wine or "tuns" that a ship could carry. The "tun" at that time was a legal standard measurement and a tun of wine was to measure not less than 252 gallons.

Over the years commissions were appointed and rules for tonnage were formulated. A Royal Commission was appointed in 1849 with a George Moorsom as its secretary and their proposals were embodied in the Merchant Shipping Act 1854. This became known as the Moorsom System on which the tonnage laws and regulations of most maritime nations were based.

The tonnage regulations were considered by another Royal Commission in 1881 and subsequently amended by further Acts up to the year 1965.

International Conference on Tonnage Measurement 1969

An International Conference on Tonnage Measurement of Ships was convened in London by the Inter-Governmental Maritime Consultative Organization (IMCO) in 1969. This conference was the first major international review of the subject.

The aims and guiding principles originally laid down by the IMCO

Sub-Committee on Tonnage Measurement were:
- a) it should not influence design, and in particular it should not encourage constructional features which detract from safety or efficiency;
- b) it should avoid dependence upon details of construction;
- c) it should permit the determination of tonnages in the early design stage of the ship and provide, insofar as possible, for the use of plans for physical measurement;
- d) it should be as direct and simple as possible, consistent with the purposes to be served;
- e) it should not adversely affect the economics of the shipping industry;
- f) it should embody a concept avoiding needless and objectionable features relative to exemptions in superstructures.

The Conference prepared the International Convention on Tonnage Measurement of Ships 1969 and at the Assembly of IMCO during October 1969 a Resolution was adopted in which the outcome of the Tonnage Measurement Conference was approved and Member States were invited to accept the Convention.

It is important to observe that gross and net tonnages—Regulations 3 and 4—are no longer expressed in the time-honoured unit of the "ton" of 100 ft^3 (2·83 m^3), the new unit being a function of cubic metres. Under the Convention a ship will be described as having "Gross Tonnage 1234" or "Net Tonnage 1234", the word "tons" no longer appears.

The new convention consists of twenty-two articles with seven annexed regulations, providing for gross and net tonnages which are computed independently. The gross tonnage is a function of the moulded volume of *all* enclosed spaces of the ship. The net tonnage is given by a formula which is a function of the moulded volume of all *cargo* spaces of the ship with corrections for draught less than 75 per cent of the depth of the ship and for the number of berthed and unberthed passengers. The net tonnage so calculated must not be less than 30 per cent of the gross tonnage.

A summary of the regulations for determining gross and net tonnages of ships under the International Convention of 1969 is given below:

Regulation 1
The tonnage of a ship shall consist of gross tonnage and net tonnage.

Regulation 2
1) *Upper Deck.* The uppermost complete deck exposed to weather and sea, which has permanent means of weathertight closing of all openings

in the weather part, and below which all openings in the sides of the ship are fitted with permanent means of watertight closing.

2) *Moulded Depth.* The vertical distance measured from the top of the keel to the underside of the upper deck at side.

3) *Breadth.* Maximum breadth of the ship, measured amidships to the moulded line of the frame.

4) *Enclosed Spaces.* are all those spaces which are bounded by the ship's hull, by fixed or portable partitions or bulkheads, by decks or coverings other than permanent or movable awnings.

5) *Excluded Spaces.* Details of the excluded spaces are given in Regulation 2 para (5) (a) to (e). Any space which fulfils at least one of the following three conditions shall be treated as an enclosed space:
 —the space is fitted with shelves or other means for securing cargo or stores;
 —the openings are fitted with any means of closure;
 —the construction provides any possibility of such openings being closed.

6) *Passenger.* Every person other than:
 a) the master and the members of the crew or other persons employed or engaged in any capacity on board a ship on the business of that ship;
 and
 b) a child under one year of age.

7) *Cargo Spaces* to be included in the computation of net tonnage are enclosed spaces appropriated for the transport of cargo which is to be discharged from the ship, provided such spaces have been included in the computation of gross tonnage. Such cargo spaces shall be certified by permanent marking with the letters CC (cargo compartment) to be so positioned that they are readily visible and not to be less than 100 mm in height.

Regulation 3. Gross Tonnage
 The gross tonnage (GT) of a ship shall be determined by the following formula:

$$GT = K_1 V$$

where:
 V = Total volume of all enclosed spaces of the ship in cubic metres
 $K_1 = 0 \cdot 2 + 0 \cdot 02 \log_{10} V$ (or as tabulated in Appendix 2).
 See Table 6.2.

Regulation 4. *Net Tonnage*
1) The net tonnage (NT) of a ship shall be determined by the following formula:

$$NT = K_2 V_C \left[\frac{4d}{3D}\right]^2 + K_3 \left[N_1 + \frac{N_2}{10}\right]$$

in which formula:

a) the factor $\left[\dfrac{4d}{3D}\right]^2$ shall not be taken as greater than unity;

b) the term $K_2 V_C \left[\dfrac{4d}{3D}\right]^2$ shall not be taken as less than $0 \cdot 25\ GT$;

and

c) NT shall not be taken as less than $0 \cdot 30\ GT$;

and in which:

V_C = total volume of cargo spaces in cubic metres
$K_2 = 0 \cdot 2 + 0 \cdot 02\ \log_{10} V_C$ (or as tabulated in Appendix 2).
See Table 6.2.
$K_3 = 1 \cdot 25 \dfrac{GT + 10{,}000}{10{,}000}$

D = moulded depth amidships in metres, as defined in Regulation 2
d = moulded draught amidships in metres, as defined in 2) of this Regulation
N_1 = number of passengers in cabins with not more than 8 berths
N_2 = number of other passengers
$N_1 + N_2$ = total number of passengers the ship is permitted to carry, as indicated in the ship's passenger certificate; when $N_1 + N_2$ is less than 13, N_1 and N_2 shall be taken as zero
GT = gross tonnage of the ship, as determined in accordance with the provisions of Regulation 3.

2) The moulded draught (d) referred to in paragraph 1) of this Regulation shall be one of the following draughts:
   a) for ships to which the International Convention on Load Lines in force applies, the draught corresponding to the summer load line (other than timber load lines);
   b) for passenger ships, the draught corresponding to the deepest subdivision load line assigned in accordance with the International Convention for the Safety of Life at Sea in force;
   c) for ships to which the International Convention on Load Lines does not apply, but which have been assigned a load line in compliance with national requirements, the draught corresponding to the summer load line so assigned;

d) for ships to which no load line has been assigned, but the draught of which is restricted in compliance with national requirements, the maximum permitted draught;
e) for other ships, 75 per cent of the moulded depth amidships, as defined in Regulation 2 (2).

TABLE 6.2.
Coefficients K_1 and K_2 referred to in Regulations 3 and 4 (1)
V or V_c = Volume in Cubic Metres

V or V_c	K_1 or K_2	V or V_c	K_1 or K_2	V or V_c	K_1 or K_2	V or V_c	K_1 or K_2
10	0·2200	45,000	0·2931	330,000	0·3104	670,000	0·3165
20	0·2260	50,000	0·2940	340,000	0·3106	680,000	0·3166
30	0·2295	55,000	0·2948	350,000	0·3109	690,000	0·3168
40	0·2320	60,000	0·2956	360,000	0·3111	700,000	0·3169
50	0·2340	65,000	0·2963	370,000	0·3114	710,000	0·3170
60	0·2356	70,000	0·2969	380,000	0·3116	720,000	0·3171
70	0·2369	75,000	0·2975	390,000	0·3118	730,000	0·3173
80	0·2381	80,000	0·2981	400,000	0·3120	740,000	0·3174
90	0·2391	85,000	0·2986	410,000	0·3123	750,000	0·3175
100	0·2400	90,000	0·2991	420,000	0·3125	760,000	0·3176
200	0·2460	95,000	0·2996	430,000	0·3127	770,000	0·3177
300	0·2495	100,000	0·3000	440,000	0·3129	780,000	0·3178
400	0·2520	110,000	0·3008	450,000	0·3131	790,000	0·3180
500	0·2540	120,000	0·3016	460,000	0·3133	800,000	0·3181
600	0·2556	130,000	0·3023	470,000	0·3134	810,000	0·3182
700	0·2569	140,000	0·3029	480,000	0·3136	820,000	0·3183
800	0·2581	150,000	0·3035	490,000	0·3138	830,000	0·3184
900	0·2591	160,000	0·3041	500,000	0·3140	840,000	0·3185
1,000	0·2600	170,000	0·3046	510,000	0·3142	850,000	0·3186
2,000	0·2660	180,000	0·3051	520,000	0·3143	860,000	0·3187
3,000	0·2695	190,000	0·3056	530,000	0·3145	870,000	0·3188
4,000	0·2720	200,000	0·3060	540,000	0·3146	880,000	0·3189
5,000	0·2740	210,000	0·3064	550,000	0·3148	890,000	0·3190
6,000	0·2756	220,000	0·3068	560,000	0·3150	900,000	0·3191
7,000	0·2769	230,000	0·3072	570,000	0·3151	910,000	0·3192
8,000	0·2781	240,000	0·3076	580,000	0·3153	920,000	0·3193
9,000	0·2791	250,000	0·3080	590,000	0·3154	930,000	0·3194
10,000	0·2800	260,000	0·3083	600,000	0·3156	940,000	0·3195
15,000	0·2835	270,000	0·3086	610,000	0·3157	950,000	0·3196
20,000	0·2860	280,000	0·3089	620,000	0·3158	960,000	0·3196
25,000	0·2880	290,000	0·3092	630,000	0·3160	970,000	0·3197
30,000	0·2895	300,000	0·3095	640,000	0·3161	980,000	0·3198
35,000	0·2909	310,000	0·3098	650,000	0·3163	990,000	0·3199
40,000	0·2920	320,000	0·3101	660,000	0·3164	1,000 000	0·3200

Coefficients K_1 or K_2 at intermediate values of V or V_c shall be obtained by linear interpolation

STATUTORY REGULATIONS

Registration

The compulsory registration of British ships was brought about by the Navigation Acts of 1660 and onwards. The Registry Act of 1786 made it compulsory for every ship to have the name of the vessel and the port to which she belonged painted on the stern and the certificate of registration had to contain details of dimensions.

By the Merchant Shipping Act of 1894 every British ship, with certain minor exemptions, must be registered. A vessel coming within the Act and not so registered is not considered a British ship. The ship's master must always have the certificate of registry—termed the ship's register—in his possession on board; in default of this the ship is liable to be detained. Prior to registration the ship must be surveyed, measured for tonnage, and the draught marks indicated on the stem and stern post. Before a vessel proceeds to sea the draughts must be recorded in the official log book and reported to the Customs Authority.

On completion of the registration survey, a Certificate of Registry is prepared by the Surveyor and forwarded to the Registrar at the intended port of registry. This Certificate indicates particulars of the build of the ship and dimensions by which the ship may be identified, also particulars of tonnage measurement and certain details of the propelling machinery. A certification is also made of the draught marks. These are cut in on each side of the stem and sternpost either in Roman capital letters or figures, not less than 152 mm in height and painted white or yellow on a dark ground.

Application for registry must be made by the owner of the ship and the application accompanied by a special declaration of ownership. For the registry of a new ship, a Builder's Certificate must also be submitted to the Registrar. When all the necessary documents are available to the Registrar, the required particulars are entered in the Official Register Book under the next available number and eventually a Certificate of Registry is prepared for the ship. This certificate is essential in connexion with most of the ship's business and movements. Prior to the actual delivery of this certificate a carving note is issued by the Registrar giving details of the required markings on the ship, these being:

    a) the ship's name to be marked on each side of the bow and the name and port of registry on the stern.

    b) the official number and net tonnage to be marked on a main beam.

The satisfactory marking of these particulars is certified by the surveyor on the carving note, and the certificate of registry can then be issued by the Registrar in exchange for the duly signed carving note.

A ship may have the name and port of registry changed under certain regulated circumstances, but the official number allotted to a ship on first registry is never changed. If a ship ceases to be a British ship by reason of sale or other circumstances, the vessel's registry book must be returned to the Registrar, at the appropriate port of registry, and the registration is duly cancelled. If this vessel again comes into British ownership, the ship may be re-registered after survey and will be allotted the original official number.

CHAPTER 7
IMCO

Inter-Governmental Maritime Consultative Organization

In 1948, the United Nations Maritime Conference at Geneva drew up a convention which created IMCO. The purposes of the new Organization were designed to cover the whole field of sea transport and to provide a means for co-operation among governments on technical matters affecting international merchant shipping, with special emphasis on the safety of life at sea. The IMCO Convention required the formal approval of 21 States before the Organization could start to function. This was achieved in March 1958 and on the 6th January 1959 the IMCO Assembly met in London. The first permanent international maritime body had come into being.

Before detailing the achievements of IMCO during the past decade it may be of interest to examine the background of sea transport—one of the oldest callings of man. Due to its essentially international character, sea transport has for ages demanded a high standard of co-operation between the maritime countries of the world, but lacked a central organization to co-ordinate activities. In spite of the extensive co-operation of governments where the saving of life at sea was concerned it was not until 1889 that the first international maritime conference took place. This Conference, held in Washington, discussed matters such as: regulations for preventing collisions at sea, saving of life and property from shipwreck, qualifications for officers and seamen, lanes for vessels on frequented routes, the establishment of a permanent international maritime commission etc. However, on the latter matter, the Conference concluded that the establishment of such a commission was not, at that time, expedient.

In 1897 the International Maritime Committee was formed to cope with the legal aspects of merchant shipping. This body also assisted in the work of several international conferences including the one that drafted the 1914 Convention for the Safety of Life at Sea. This latter was a direct result of the loss of the *Titanic* in 1912. The 1914 Convention never became fully operative because of the First World War. The United Kingdom Government proposed a further Conference to prepare up-to-date requirements and this proposal culminated in the London Conference of 1929. This ultimately produced the Convention for the Safety of Life at Sea, 1929. In 1930 another international conference drew up regulations determining the load line of merchant ships engaged in international trade.

In 1945 a great advance in inter-governmental co-operation was made by the creation of the United Nations. As stated above, the United Nations Maritime Conference at Geneva in 1948 drew up the convention which created IMCO.

International Conventions

All but two of the undernoted conventions (a) and (c) were the result of conferences called by IMCO. The list, however, gives a clear indication of the activities of the Organization carried out in addition to its statutory functions.
(a) International Convention for the Safety of Life at Sea, 1948.
(b) International Convention for the Safety of Life at Sea, 1960.
(c) International Convention for the Prevention of Pollution of the Sea by Oil, 1954 as amended in 1962.
(d) Convention on Facilitation of International Maritime Traffic, 1965.
(e) International Convention on Load Lines, 1966.
(f) International Convention on Tonnage Measurement of Ships, 1969.
(g) International Convention relating to Intervention on the High Seas in cases of Oil Pollution Casualties, 1969.
(h) International Convention on Civil Liability for Oil Pollution Damage, 1969.

Safety of Navigation

A great deal of effort has been made by IMCO in the direction of introducing measures and policies designed to increase the safety of navigation. Among the most important are those which concern the compulsory carriage of navigational equipment and the principle of ships' routeing and separation of traffic at sea.

Navigational equipment—such as radar, echo-sounders, gyro-compass and direction finders—which have to the present time been carried at the discretion of the owner, will now be made mandatory in ships above a certain size.

An attempt is now being made to lay down for the first time an international set of performance specifications and testing procedures for radar, gyro-compasses and echo-sounding devices.

Considerable effort has been concentrated on two main aspects of safety of navigation—measures for regulating traffic in areas of high density or converging routes and revision of the International Regulations for Preventing Collisions at Sea. Traffic separation schemes have been established in some 50 areas where there is dense or converging traffic. The object of any such scheme is to reduce the number of ships meeting on opposite, or nearly opposite courses and thus to reduce the risk of collision. Detailed descriptions of the schemes are included in national maritime publications, charts etc., and IMCO has issued a comprehensive booklet on the subject.

In addition to these measures, IMCO has made a number of recommendations to governments concerning the provision of pilotage and port advisory services, electronic position-fixing equipment and identification lights for deep-draught ships in narrow channels.

An important current item is the revision of the International Regulations for Preventing Collisions at Sea which when completed, will become the new mariners' code. Among other things, this revision includes measures for improving the efficiency of navigation lights and sound signals.

Another important subject under review is the use of space techniques for determination of ships' positions.

IMCO is taking an active part in the United Nations programme on exploration and exploitation of the world oceans. The Organization is considering various safety aspects of operating scientific stations in the marine environment.

Radio Communications

A wide range of operational questions designed to improve or reshape the existing Maritime Distress System is being studied. The use of space communication techniques for shipping is also being considered and a list of operational requirements has been prepared and sent out to governments in preparation for the Space Conference to be convened under the auspices of the International Telecommunication Union.

Amendments have been made to the 1960 Safety Convention affecting radio communications to bring the Convention into line with technical developments as reflected in the amended Radio Regulations.

The ever-increasing use of mobile drilling units has created a need for determining the radio-communication requirements for such units so that they can be integrated into the existing communication system not only for operational purposes but also in cases of distress.

Radio-communication facilities for the ever increasing use of the newer types of craft such as hydrofoil boats and air-cushion vehicles are under consideration.

Life Saving Appliances

IMCO has developed standards for the testing and approval of life-jackets, requirements concerning life-saving appliances for air-cushion vehicles and for mobile off-shore units engaged in the exploration and exploitation of resources of the sea bed and ocean floor.

When current studies are completed, the standards established for the novel types of craft mentioned above, and for the mobile units will eventually form part of a comprehensive code covering all aspects of the safety of such craft.

Studies are being carried out on the feasibility of amending certain regulations of the 1960 Safety Convention referring to life-saving appliances for passenger ships on short international voyages, while the provision of special fire-resisting lifeboats for tankers is also under consideration.

Search and Rescue Manual

As a guide for masters and others involved in distress incidents at sea, IMCO has prepared the Merchant Ship Search and Rescue Manual (MERSAR). It contains specific instructions on the actions to be taken by the vessel in distress, by those assisting or participating in the search, and general guidelines on the organization and conduct of such search and rescue operations.

Training Guidance

The training of masters, officers and seamen is essentially an integral part of safety at sea. The joint ILO/IMCO Committee on Training has issued the "Document for Guidance, 1970". It includes syllabuses on various subjects for inclusion in maritime training programmes.

Subdivision and Stability of Ships

A proposal for new subdivision regulations for passenger ships is being developed based on the method of probability concerning ship survival. This proposal which also takes into account the longitudinal subdivision of passenger ships of the ferry type is to replace the existing requirements in the 1960 Safety Convention.

Following adoption of the Recommendation on Intact Stability of Passenger and Cargo Ships under 100 m. IMCO has under consideration a basis for evelution of improved stability criteria.

For ships carrying timber deck cargo, stability requirements are being developed.

Safety of Fishing Vessels

Since the adoption of the Recommendations on Intact Stability of Fishing Vessels, a further study is in progress on simplified stability criteria intended for judging the stability of small fishing vessels for which no stability data is available.

Also being considered is the matter of minimum freeboards for fishing vessels, together with those aspects of the construction of such vessels that affect the vessel's stability and the safety of the crew.

IMCO is now preparing Part B of a Code of Safety for Fishermen and Fishing Vessels;
Part A is Safety and Health Practice for Skippers and Crews.
Part B Safety and Health Requirements for the Construction and Equipment of Fishing Vessels.

Tanker Construction and Equipment

Studies of the construction and equipment of oil tankers from the viewpoint of preventing or limiting pollution of the sea by oil in the event of stranding or collision were first taken up at the end of 1968. The studies not only embraced considerations such as the probability of oil outflow in the event of damage to tankers but also comprehensive investigations into the economic implications of tank size limitation.

In order to limit the amount of oil outflow in the event of collisions or groundings of large tankers, studies are being carried out on several factors such as, the fitting of double bottoms, the arrangement of tanks, the interposing of clean water ballast tanks etc.

IMCO's Maritime Safety Committee made a recommendation to governments that the maximum tank size of the largest tankers should be limited to 50,000 m^3 for centre tanks and 30,000 m^3 for wing tanks. Based on proposals by one of its technical sub-committees the Maritime Safety Committee decided on definite arrangements and limitation of tank size, these to be incorporated as amendments to the International Convention on Prevention of Pollution of the Sea by Oil 1954. The new requirements limit the hypothetical oil outflow in the event of collision or stranding of oil tankers to a value of 30,000 m^3. In the case of normal tankers with two longitudinal bulkheads, the capacity of a single centre tank and a wing tank will be limited to 30,000 m^3 and 15,000 m^3 respectively.

Chemical Bulk Carriers

In view of the increase in the sea transportation of hazardous or noxious chemical products in bulk it became apparent that there was need for international measures to ensure their safe carriage: the Maritime Safety Committee approved an interim recommendation for existing ships of the tanker type carrying dangerous chemicals in bulk in liquid form.

A code for new ships has now been approved and will be developed further in the light of the rapid development in this form of transport.

Fire Safety in Ships

Fire is one of the most serious hazards of ships. IMCO recommended a series of appropriate amendments to the 1960 Safety Convention for existing passenger ships and a number of amendments which would apply to new passenger ships only. Recommendations have been made on fire safety measures for hydrofoil boats.

A study is being made on fire safety of ships under automated control together with the fire protection of cargo ships.

Fire safety of oil tankers is a matter of universal concern in view of the rapid increase in size and number of these ships. Requirements for structural fire protection and fire extinguishing equipment for new tankers are being developed which will lead to amendments to the existing Safety Convention. Work is in progress on studies of explosion hazards in large tankers.

Carriage of Dangerous Goods

The International Maritime Dangerous Goods Code has now been adopted by a number of governments and the code concerning the carriage of explosives has been completed.

The carriage of dangerous goods in limited quantities, or in packages or drums is under consideration as well as the carriage of these goods in roll-on/roll-off ships.

Carriage of Bulk Cargoes

Detailed recommendations on the carriage of grain in bulk including regulations for the strength of grain fittings have been approved by IMCO.

A Code gives advice on general precautions to be taken with bulk shipments of ores, ore concentrates and similar cargoes. The types of cargoes covered are listed separately in a series of appendices and periodic revision keeps the Code up-to-date.

Container Transport

Container transport and the safety aspect have been under review for some time. The subject was originally raised by IMCO in 1968. The object was to produce proposals which would be suitable for international agreement. The proposals deal with such container problems as strength, construction, marking, testing and certification, special types of containers, safe handling and stowage and securing on board ships.

It is proposed to hold a conference on international container traffic.

Marine Pollution

An international conference convened at the invitation of the Government of the United Kingdom resulted in the International Convention for the Prevention of Pollution of the Sea by Oil, 1954.

This was deposited with that government and transferred to IMCO when it was established in 1959.

The 1954 convention deals only with the deliberate or operational discharge of oil from ships and does not relate to pollution arising from maritime accidents.

Extensive amendments to the 1954 convention were adopted in 1969 and these may be summarized as follows:
   a) prohibition of deliberate discharge
   b) prevention of accidental discharge
   c) powers to States for dealing with pollution
   d) provision for redress for damage caused, and
   e) methods for dealing with spillages.

At the 1973 IMCO Conference on Marine Pollution the main objective of the conference is the complete elimination of wilful and intentional marine pollution by oil and other noxious pollutants and minimization of accidental spills.

Ship Losses

In spite of modern ship construction, advanced navigational equipment and elaborate communication systems, 70 ships have, in the past decade, disappeared without trace. These vessels have left no radioed SOS, no survivors and no recognizable wreckage.

Some of the 70 ships were fishing vessels and it has been suggested as a possible explanation that they capsized and sank in heavy seas. But larger ships have disappeared such as
   a) the *Milton Iatridis* of 10,000 tons. This ship left New Orleans for Cape Town with a cargo of vegetable oil and caustic soda. Instructions had been given to radio every four or five days. No message was received and finally Lloyd's formally posted her missing.
   b) the *Ithaca Island* of 7,400 tons left Norfolk, Virginia for Manchester with a cargo of grain. No distress signals were received and an air search found no wreckage. Finally the ship was posted as missing.
   c) the *Iligan Bay* left the Continent for Manila. Off Corunna she radioed that she was in heavy seas. That was the last heard of the ship. Two months after she should have arrived in Manila the vessel was posted as missing by Lloyd's.
   d) the *Kiki* a 3,750 ton ship sailed from Emden in Germany for Dugirat in Yugoslavia. Fourteen days later a Dutch radio station heard her call sign but no position was given. That was all. No survivors, no bodies and in spite of the very large number of ships that cross the North Sea no definitive wreckage was observed.

Ship casualties are reported under the following categories: foundered, missing, burnt, collision, wrecked, and lost.

Between 1961 and 1969 inclusive Lloyd's posted 2,520 ships as lost, and these were classified thus:

Foundered	676
Missing	63
Burnt	365
Collision	277
Wrecked	1,063
Lost	76
Total lost	2,520

Lloyd's Register of Shipping issue a quarterly casualty return giving details of merchant ships totally lost, broken up, etc.

PART II

NAVAL ARCHITECTURE

The subject of Naval Architecture is very old, dating back to Noah and before. In the early days, it was founded on empiricism with a scientific veneer. Today it is much more scientific but still leaning slightly on empiricism. During the past four decades great changes have taken place in Naval Architecture. The science has broadened and ranges from fluid dynamics to the theory of probabilities with numerous ancillary branches. In the past progress has been made, for the most part, by a gradual process of trial and elimination.

Naval Architecture is a living and growing science. The amount of scientific knowledge is growing exponentially and it is essential to remember that although generalities may be important the basic principles are vital. The principles are outlined in the Chapters that follow.

CHAPTER 8
Definitions of Principal Terms used in Naval Architecture

Naval Architecture like all subjects has terms unique to itself. It is convenient to have a terminology and a shorthand in the form of abbreviations and symbols.

It is essential in the study of the subject to have a clear understanding of the meaning of the various terms, abbreviations and symbols used.

Forward Perpendicular (FP) is a vertical line through the intersection of the load waterline and the stem contour. Figure 8-1.

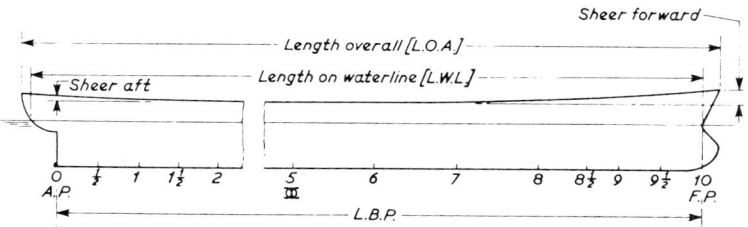

FIG. 8-1—*Illustration of terms used in naval achitecture.*

After Perpendicular (AP) is at where the aft side of the sternpost meets the load waterline *or* if there is no sternpost at the centre of the rudder stock.

Length between Perpendiculars (LBP) is the horizontal distance between the forward and after perpendiculars.

Amidships (⊗) is the midway point between the perpendiculars.

Midship Section is the transverse section of the ship at amidships.

Breadth Moulded (B mld) is measured at amidship and is the maximum breadth over the frames. Figure 8-2.

Depth Moulded (D mld) is the vertical distance at amidships from top of keel to the top of the deck beam at side *or* underside of deck plating at the ship side: the deck to which the depth is measured should be stated.

Draught Moulded (H mld) is the distance of the top of the keel below the waterline.

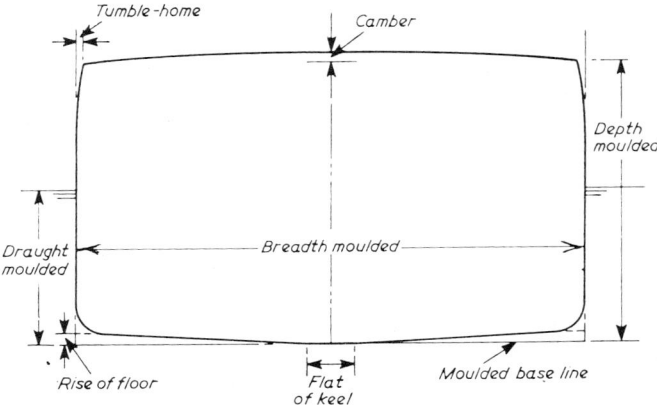

FIG. 8-2—*Illustration of terms used in naval architecture.*

Draught (H) is the distance of the lowest point of the keel below the waterline.

Moulded Base Line. This is a horizontal line which passes through the top of the keel at amidships. It acts as the datum or base line for all hydrostatic calculations.

Sheer. The longitudinal curvature given to decks. It is measured by the difference in height of side at any point and the height at amidships. The amount of sheer at the FP is often twice the sheer at the AP.

Camber. This is the curvature given to a deck transversely. It is measured by the difference between the heights of the deck at side and centre. The camber amidships is frequently one-fiftieth of the breadth (B) of the ship.

Rise of Floor is the rise of the bottom shell plating measured transversely amidships at the moulded breadth line.

Tumble-Home. The fall in of the sides amidships.

Flare. The outward curvature of the forward sections above the waterline.

Rake. The departure from the vertical of any line in profile such as the stem.

Trim. The longitudinal inclination of a ship is measured by the difference between the draughts forward and aft. When the draughts are the same forward and aft the ship is said to be "on an even keel". "Down by the head" when the draught forward is greater than that aft. "Down by the stern" when the draught aft is greater.

Heel. The inclination of the ship in the transverse direction, measured in degrees.

Yaw. The movement from the mean course of a ship in the horizontal plane, measured in degrees.

DEFINITIONS OF PRINCIPAL TERMS USED IN NAVAL ARCHITECTURE 113

Middle Line Plane. Ships have only one plane of symmetry, called the middle line plane and it is the principal plane of reference. The shape of the ship given by this plane is known as the profile.

Waterplanes are planes at right angles to the middle line plane; they are symmetrical about the middle line plane. Waterplanes looked at edge on in the profile are called waterlines.

Transverse Planes perpendicular to the middle line plane show the shape of vertical sections of the ship.

Freeboard. The vertical distance between the actual or permissible waterline and the upper surface at side of the deck to which it is to be measured.

Load Line Mark. All merchant ships, with a few exceptions, must be marked with a load line. The upper edge of this line indicates the maximum permissible draught. The load lines are set off amidships, on both sides of the ship, at specified distances below a deck line. The standard markings for a cargo ship are as under:

S = Summer   T = Tropical
W = Winter   F = Fresh Water
          WNA = Winter North Atlantic
          TF = Tropical Fresh Water

The upper edge of the summer line, if continued, passes through the centre of the load line disc and is the basic line.

Further details are given in Chapter 6.

Parallel Middle Body. The length over which the midship section remains constant in area and shape. Figure 8-3.

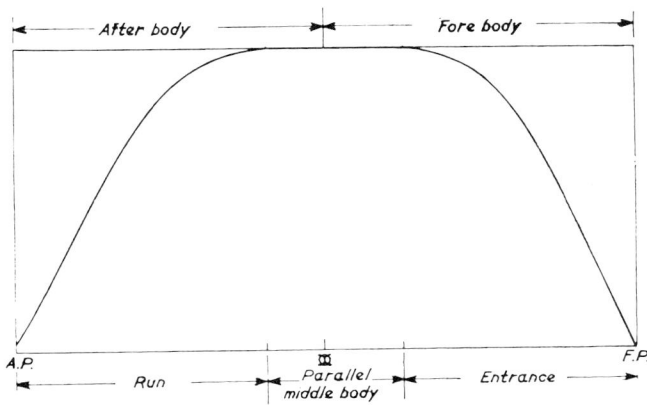

FIG. 8-3—*Parallel middle body.*

Fore Body. The immersed body forward of the midship section.
After Body. The immersed body aft of the midship section.

Entrance. The immersed body forward of the parallel body.
Run. The immersed body aft of the parallel body.
Lines Plan. The delineation of the ship's form in three views:
1) The longitudinal elevation or profile which gives the general outline of the ship, the position and sheer of the decks.
2) The half-breadth plan shows the shape of the decks and the waterlines which are formed by the intersection of the surface of the ship with horizontal planes.
3) The body plan which shows the transverse sections. The lines plan represents the moulded surface of the ship, that is the inside of shell plating. The accepted convention for the profile is to show the bow pointing to the right.

Since both sides of the ship are identical, only one is delineated in the Body Plan and Half Breadth Plan. On the Body Plan the fore body is drawn to the right of the centre line and the after body to the left. Figure 8–4 shows the Body Plan of a cargo ship.

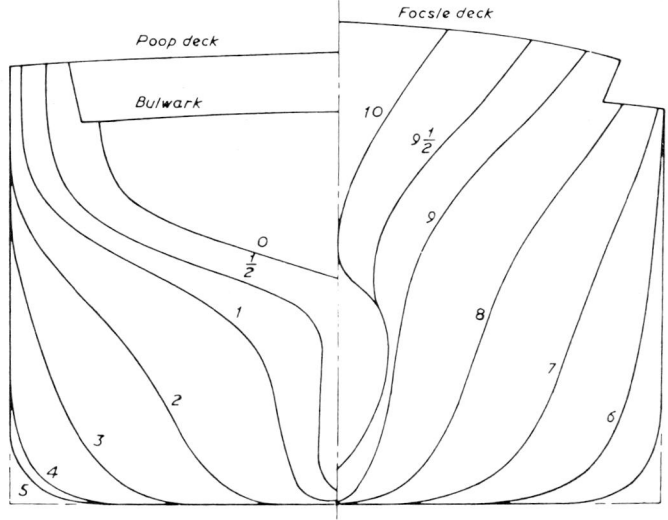

FIG. 8–4—*Body plan of cargo ship.*

For design purposes the profile between the perpendiculars—the AP and FP—is divided into 10 equal parts. These stations are conventionally numbered from aft to forward; the AP is N⁰ 0 and the FP is N⁰ 10, with half-stations at the ends where the form changes rapidly. Figure 8–1. Transverse sections at these stations are seen as curved lines in the body plan.

The profile and body are also divided vertically by equi-spaced planes parallel to the base line. These planes looked at edge on in the profile and body appear as straight lines and are called waterlines. In the half-breadth they appear as curves.

Centre of Flotation (CF) is the centroid of the waterplane area. For small angles of trim consecutive waterlines pass through the CF.

Centre of Buoyancy (B) is the centroid of the underwater volume of the ship and is the point through which the total force of buoyancy can be assumed to act. Its position is defined thus:
    a) KB = the vertical distance above base.
    b) LCB = the longitudinal distance from amidships.

Centre of Gravity or Centre of Mass (G) is the point through which the total mass of the ship may be assumed to act. The position is defined thus:
    a) KG = the vertical distance above base.
    b) LCG = the longitudinal distance from amidships.

Tonnes per Centimetre (TPC) of a waterplane is the mass required to effect a change in the mean draught at that waterline of 1 cm.

Light Weight or Light Mass. This is the displacement of the ship when complete and ready for sea but no crew, passengers, baggage, stores, fuel, water or cargo on board. Boilers, if any, are filled with water to working level.

Deadweight or Dead Load. This is the difference between the displacement at any draught, and the light mass.

Total deadweight or dead load is thus the difference between the load displacement and the light mass.

The *load* displacement is that at the maximum permissible draught.

The deadweight includes crew, passengers, baggage, stores, fuel, water and cargo.

Thus *displacement = light mass + deadweight or deadload*.

Deadweight Coefficient (C_D). A coefficient used in the early stages of estimating design displacement and dimensions.

$$C_D = \frac{\text{Total deadweight}}{\text{Load displacement}}$$

So that for a proposed ship of 7,000 tonnes total deadweight with say a deadweight coefficient of 0·70 the approximate displacement is given by

Load displacement $= \dfrac{\text{Total deadweight}}{C_D} = \dfrac{7000}{0 \cdot 7} = 10,000$ tonnes

Deadweight or Deadload Scale

    The shipowner, master and officers are concerned with deadweight or deadload. This is presented in tabular form as a deadweight scale against a vertical scale of draught. A typical deadweight scale is shown in Figure 8–5. This generally includes a scale of tonnes per centimetre (TPC) in view of its importance to masters and officers.

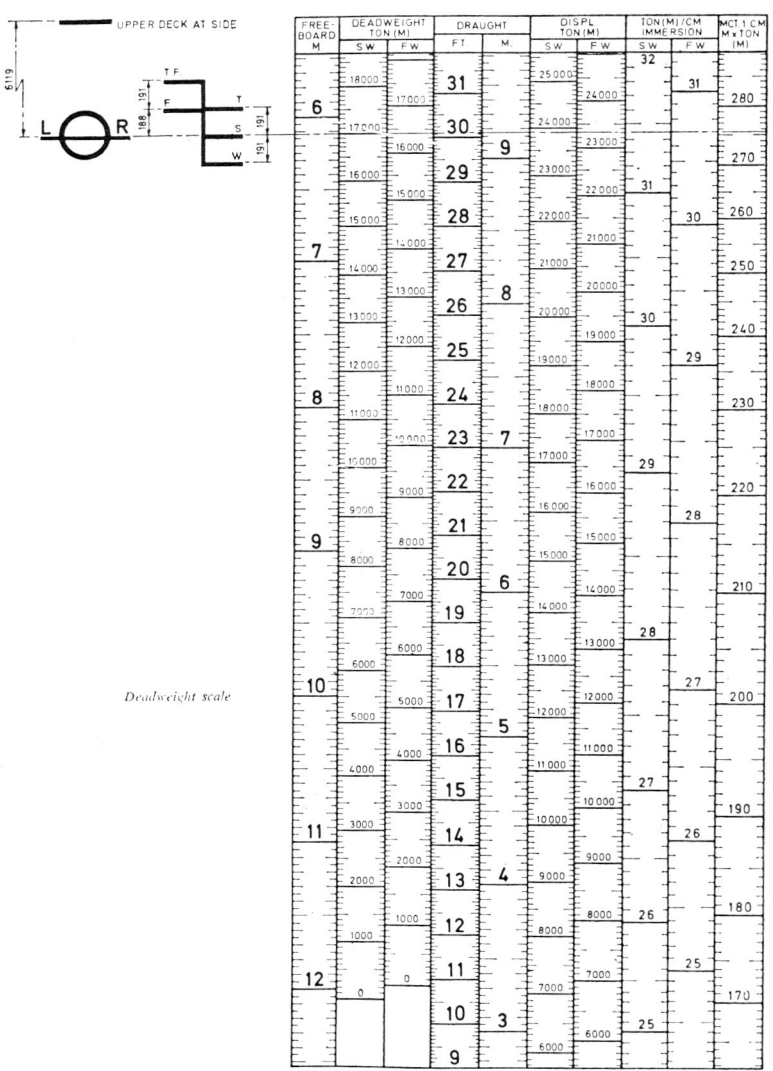

FIG. 8–5—*Deadweight scale.*

The approximate load displacement of a ship can be determined **from the deadweight scale as follows:**

Determine the TPC at the load draught from the scale. From Figure 8-5 say in this case 31·5.

Note the sinkage permitted for fresh water as indicated on the load line mark say 19 cm.

Then approximate load displacement is given by

TPC × FW sinkage × 40 = 31·5 × 19 × 40
= 23,940 tonnes

Form Coefficients
  Block, Prismatic, Midship Area, Waterplane Area
These are dealt with in Chapter 10.

Bonjean Curves. These are curves of transverse sectional area drawn against a vertical scale of draught, thus the ordinate at say the load waterline (LWL) for a particular station gives to some scale the area of the section up to the LWL Figure 8-6.

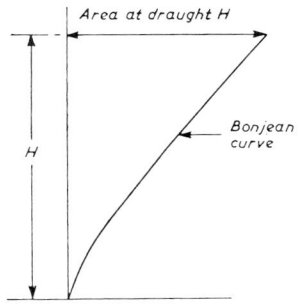

FIG. 8-6—*Bonjean curves.*

Scantlings. The dimensions and the thickness of rolled sections and the breadth and thickness of plates which together compose the ship's structure or part of same.

CHAPTER 9
SI Units
International System of Units

SI is the accepted abbreviation for the International System of Units agreed at the General Conference of Weights and Measures in 1960. The system is an extension and refinement of the traditional metric system. A decimal system of units was introduced by Simon Stevin (1548–1620) and the metric system was legally adopted in France in 1795. This system was based on the metre as the unit of length and the gramme as the unit of mass.

Table 9.1

Quantity	Unit	Unit Symbol
Length	Metre	m
Mass	Kilogramme	kg
Time	Second	s
Force	Newton	$N = \text{kg m/s}^2$
Work	Joule	$J = \text{N m}$
Power	Watt	$W = \text{J/s}$

Prefixes to denote multiples and sub-multiples to be affixed to the names of units are:

Factor by which unit is multiplied	Prefix	Symbol
$1{,}000{,}000 = 10^6$	Mega	M
$1{,}000 = 10^3$	Kilo	k
$100 = 10^2$	Hecto	h
$10 = 10^1$	Deca	da
$0 \cdot 1 = 10^{-1}$	Deci	d
$0 \cdot 01 = 10^{-2}$	Centi	c
$0 \cdot 001 = 10^{-3}$	Milli	m

In 1873 the British Association for the Advancement of Science selected the centimetre, the gramme and the second as basic units of length, mass and time for scientific purposes, thus the cgs system.

The SI metric system is now coming into world use; about thirty countries have decided to make SI the only legally accepted system, and in this country the shipbuilding industry is in process of making the changeover.

There are six primary units in the SI system and these are: length, mass, time, electric current, temperature and luminous intensity. In naval architecture the main concern is with length, mass and time and the quantities which can be directly derived from them. Some of these are indicated in Table 9.1.

Table 9.2

Quantity	SI Unit	Unit Symbol
Area	Square Metre	m^2
Volume	Cubic Metre	m^3
Density	Kilogramme per Cubic Metre	kg/m^3
Velocity	Metre per Second	m/s
Acceleration	Metre per Second Squared	m/s^2
Pressure: Stress	Newton per Square Metre	N/m^2

Quantity	Imperial Unit	Equivalent SI Unit
Length	1 ft	0·3048 m
	1 nautical mile international	1,852 m
Area	1 ft^2	0·0929 m^2
Volume	1 ft^3	0·02832 m^3
Velocity	1 ft/s	0·3048 m/s
	1 knot (internat.)	0·5144 m/s
Standard Acceleration g	32·174 ft/s^2	9·8066 m/s^2
Mass	1 lb	0·4536 kg
	1 ton	1,016·05 kg
Force	1 lbf	4·4482 N
Pressure	1 lb/in^2	6,894·76 N/m^2
Stress	1 ton/in^2	15·444 × 10^6 N/m^2
Energy	1 ft lb	1·3558 J
Power	1 hp	745·7 W
Density SW	64 lb/ft^3	1·025 $tonne/m^3$
	35 ft^3/ton	0·975 $m^3/tonne$
FW	62·2 lb/ft^3	1·000 $tonne/m^3$
	36 ft^3/ton	1·000 $m^3/tonne$
Young's Modulus (E)	13,500 $tons/in^2$	20·9 MN/cm^2

Difficulties have arisen in connexion with quantities involving loads or weights, displacement and force. It is now accepted that loads, such as cargo, and displacement be expressed in tonnes; 1 tonne = 1,000 kg. The SI unit of force is termed the Newton (Symbol N) to commemorate the name of Sir Isaac Newton and is defined as the force which, when applied to a mass of 1 kilogramme, gives it an acceleration of one metre per second squared—or 1 m/s^2.

SI UNITS — INTERNATIONAL SYSTEM OF UNITS

The SI unit of power is the watt, namely 1 joule per second. The joule is defined as the work done when a force of one newton is exerted through a distance of one metre in the direction of the force. In practice the watt is often inconveniently small and consequently the kilowatt (kW) is generally used, the kilowatt being 1,000 watts. For large powers the megawatt (MW) is used. The traditional terms BHP, SHP, DHP and EHP have been replaced by the standard nomenclature P_B, P_S, P_D and P_E. The British horse power is 745·7 W.

In Table 9·2 are given some derived units and the equivalent values of some Imperial units.

The following gives some of the quantities used in hydrostatics, hydrodynamics and propulsion, together with the expressions for deriving numerical values:

Displacement (Δ) in Tonnes (SW)
$$= \text{Immersed Volume in } m^3 \times 1·025$$
$$\text{or} = \frac{m^3}{0·975}$$

Tonnes per Centimetre (TPC)
$$= \text{Area of waterplane } (A) \text{ in } m^2 \times \frac{1}{100} \times \frac{1}{0·975}$$
$$= \frac{A}{97·5}$$

Area of Waterplane (m^2)
$$= TPC \times 97·5$$

To derive displacement from TPC values using Simpsons First Rule
$$\Delta = \text{function for volume} \times \tfrac{1}{3} \text{ CI} \times 97·5/0·975$$
$$= \text{function for volume} \times \tfrac{1}{3} \text{ CI} \times 100 \text{ tonnes}$$

Moment to Change Trim One cm (MCT 1 cm): tonne metre (tm)
An approximation to MCT 1 cm is given by $\Delta/100$ tonne metre

Wetted Surface Area (S) in m^2
An approximation to this is given by:
$$S = 2·58 \sqrt{\Delta L} \quad \Delta = \text{tonnes}; L = \text{metre}$$

Pressure is expressed in newtons per m^2
As an example: $P = HA\,w$
$$= H \times 1 \times w \times 9·81 \text{ newtons/m}^2$$
where H = head in metres
$A = 1$ m^2 $w = 1,025$ Kg/m^3 for sea water

Rudders
For middle line rudders behind single screws the force (Q) on the rudder can be taken as

$Q = 18.0 \, AV^2\theta$ newtons $A = m^2$; $V = m/s$
                                      $\theta = $ degrees

Torque $= \pi/16 \, fd^3$; $d = $ dia in metre;
 for cast steel $f = 77.22 \times 10^6 \, N/m^2$
 $= \pi/16 \times 77.22 \times 10^6 \times d^3$
 $= $ newton metre

Resistance
The Froude expression $Rf = fSV^{1.825}$
In metric units $f = $ English unit $\times 160.9$
                       $s = $ wetted surface area in m^2
                       $V = $ speed in m/s; $Rf = $ newtons

Power
$P_E = Rt \times V \times 0.5144$ $Rt = $ kilonewtons; $V = $ knots; $P_E = $ kW

$P_T = T \times Va$ $T = $ kilonewtons; $Va = $ m/sec; $P_T = $ kW

$P_D = \dfrac{2\pi QN}{1{,}000}$ $Q = $ newton metre; $N = $ rev/sec; $P_D = $ kW

$\textcircled{C} = \dfrac{579.7 \times P_E}{\Delta^{2/3} \times V^3}$ $P_E = $ kW; $\Delta = $ tonnes; $V = $ knots

Propellers
Apparent slip $= \dfrac{PN - 30.864 \, V}{PN}$ $P = $ pitch in metre; $V = $ knots
                                                  $N = $ rev/min

Real slip $= \dfrac{PN - 30.864 \, Va}{PN}$ $Va = $ knots

$Va = \dfrac{V}{1 + w}$ $w = $ Froude wake fraction

$Bp = 0.2198 \, \dfrac{NP_D^{1/2}}{Va^{2.5}}$ $N = $ rev/min; $P_D = $ kW; $Va = $ m/sec

$\delta = 3.2808 \, \dfrac{ND}{Va}$ $N = $ rev/min; $D = $ dia in metre; $Va = $ knots

Froude Circular Notation

$\textcircled{L} = 0.5824 \, \dfrac{V}{\sqrt{L}}$ V in knots

$\textcircled{M} = 1.0083 \, \dfrac{L}{\Delta^{1/3}}$ L in metres

$\textcircled{K} = 0.5848 \, \dfrac{V}{\Delta^{1/6}}$ Δ in tonnes

$\textcircled{S} = 1.0167 \, \dfrac{S}{\Delta^{2/3}}$ S in (metres)^2

SI UNITS — INTERNATIONAL SYSTEM OF UNITS

Some other relationships are given in Table 9·3.

Table 9.3

Imperial Unit	Equivalent SI Unit
1 in² ft²	0·5993 cm² m²
1 in⁴	41·62 cm⁴
1 ft⁴	0·008631 m⁴
1 ton f	9·964 kN
1 ton	1·01605 tonne
1 lb/hp/hr	0·6083 kg/kW/hr
Tonnes/in	TPI × 1·016
Tonnes/cm	$\dfrac{TPI \times 1 \cdot 016}{2 \cdot 54} = 0 \cdot 40 \times TPI$
MTI tons ft/in	$1 \cdot 016 \times \dfrac{1}{3 \cdot 28} \times \dfrac{1}{2 \cdot 54}$ tonne metre/cm
	$= MTI \times 0 \cdot 1219$ tonne metre/cm

Table 9.4—Symbols and Abbreviations

Quantity	Abbreviation	Symbol	Preferred Unit	
Length in general		L	metre	m
Length between perpendiculars	LBP		metre	m
Length on load waterline	LWL		metre	m
Length overall	LOA		metre	m
Breadth (moulded	Bmld; B ext	B	metre	m
Depth or	Dmld; D ext	D	metre	m
Draught extreme)	Hmld; H ext	H	metre	m
Amidships		ⓧ		
Displacement (as mass)		Δ	tonne	t
Displacement (as volume)		V	cubic metre	m³
Tonne per cm immersion	TPC			
Moment to change trim one centimetre	MCT 1 cm		tonne metre	tm
Wetted surface area		S	square metre	m²
Area of immersed midship section		Am	square metre	m²
Area of waterplane		Aw	square metre	m²

Quantity	Abbreviation	Symbol	Preferred Unit	
Centre of buoyancy	CB	B		
Vertical centre of buoyancy	VCB	KB		
Longitudinal centre of buoyancy	LCB			
Longitudinal centre of flotation	LCF	F		
Centre of lateral resistance	CLR			
Centre of gravity (ship)		G		
Centre of gravity (item on ship)		g		
Vertical centre of gravity (ship)	VCG			
Vertical centre of gravity (item on ship)	VCg			
Longitudinal centre of gravity	LCG			
Centre of buoyancy above keel		KB	metre	m
Centre of gravity above keel (ship)		KG	metre	m
Centre of gravity above keel (item on ship)		Kg	metre	m
Transverse metacentre above keel		KM	metre	m
Longitudinal metacentre above keel		KM_L	metre	m
Transverse metacentre above centre of buoyancy		BM	metre	m
Longitudinal metacentre above centre of buoyancy		BM_L	metre	m
Transverse metacentric height		GM	metre	m
Longitudinal metacentric height		GM_L	metre	m
Second moment of area of waterplane about centreline		I	metre^4	m^4
Second moment of area of waterplane about CF		I_{CF}	metre^4	m^4
Coefficient block		C_b		
Coefficient midship section area		C_m		
Coefficient prismatic		C_p		
Coefficient vertical		C_{vp}		
,, waterplane area		C_w		
Density		ρ	kg per m^3	kg/m^3

SI UNITS — INTERNATIONAL SYSTEM OF UNITS

Quantity	Abbreviation	Symbol	Preferred Unit	
Permeability		μ		
Resistance total		R_t	newton or kilonewton	N or kN
,, frictional		R_f		
,, residual		R_r		
,, wave making		R_w		
Scale ratio		λ		
Speed of ship		V	knot (international)	knot
		v	metre per sec	m/s
Speed of advance of propeller		Va	knot	knot
Coefficient of kinematic viscosity		va	metre per sec	m/s
		ν	centistokes	cSt
Froude number $\dfrac{V}{\sqrt{gL}}$		F_n		
Reynolds number $\dfrac{LV}{\nu}$		R_n		
Advance number $\dfrac{va}{nD}$		J		
Diameter of propeller		D	metre	m
Pitch mean effective		Pe	metre	m
Pitch face		Pf	metre	m
Revolutions per minute	rev/min	N		
Revolutions per second	rev/s	n		
Developed blade area outside boss		Ad	square metre	m²
Developed blade area including boss		Ab	square metre	m²
Blade area ratio $= \dfrac{Ad}{\pi/4 D^2}$	BAR			
Disc area ratio $= \dfrac{Ab}{\pi/4 D^2}$	DAR			
Wake fraction $\begin{cases}\text{Froude}\\ \text{Taylor}\end{cases}$		w_f w_t		
Apparent slip per cent		Sa		
Real slip per cent		Sr		
Thrust		T	kilonewton	kN
Thrust deduction fraction		t		
Power brake		P_B	kilowatt	kW
delivered		P_d	kilowatt	kW
effective		P_E	kilowatt	kW
indicated		P_I	kilowatt	kW
shaft		P_S	kilowatt	kW
thrust		P_T	kilowatt	kW
Quasi propulsive coefficient	QPC			

CHAPTER 10

Areas, Volumes, Moments, Displacement

Areas

Most calculations in naval architecture require at some stage the mensuration of an area, the moment of an area and the second moment of area. The area is generally bounded by straight lines and a curve which is usually fair and smooth. However, the curve is not of such a shape that the equation expressing it is easily determined and thus it is not possible to determine the area by mathematical integration. On the assumption that the bounding curves are portions of parabolas then approximate arithmetical rules have been devised to perform the integration.

The methods in common use are given below and summarized in Table 10.1.

Table 10.1

Rules for Determining Areas

Length of base $= L$; No. of ordinates $= N$; End ordinates are Y_1 and Y_n; No. of intervals $= N - 1$; Common interval $= h = L/(N - 1)$ except for Mean ordinate Rule when $h = L/N$

No.	Rule	Area by Rule [A]	N
1	Trapezoidal	$A = h[\frac{1}{2}Y_1 + Y_2 + Y_3 + Y_4 + Y_5 + \ldots + \frac{1}{2}Y_n]$	Odd or even
2	Mean ordinate	$A = h[Y_{1\frac{1}{2}} + Y_{2\frac{1}{2}} + Y_{3\frac{1}{2}} + \ldots + Y_{n-\frac{1}{2}}]$	Odd or even
3	Simpson's 1st Rule	$A = h/3[Y_1 + 4Y_2 + 2Y_3 + 4Y_4 + 2Y_5 + \ldots + Y_n]$	Odd
4	Simpson's 2nd Rule	$A = \frac{3}{8}h[Y_1 + 3Y_2 + 3Y_3 + 2Y_4 + 3Y_5 + \ldots + Y_n]$	4.7.10 etc.
5	$5 - 8 - 1$ Rule	$A = 1/12h[5Y_1 + 8Y_2 - Y_3]$	3

The Trapezoidal Rule

A trapezoid is a plane four-sided figure having two sides parallel, Figure 10–1. If the lengths of the parallel sides are Y_1 and Y_2 and they

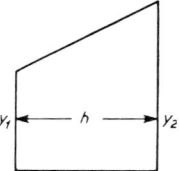

Fig. 10–1—*Trapezoid.*

are h apart then the area of the trapezoid is given by $A = h/2\,(Y_1 + Y_2)$.

The trapezoidal rule assumes that the portion of curve lying between any two consecutive ordinates Y_1 Y_2 etc., can be replaced with sufficient accuracy by a straight line. A curvilinear figure can be divided into a number of approximate trapezoids by covering it with N equally spaced vertical ordinates, h apart, the ordinates in order being Y_1 Y_2, Y_3 ...... Y_n Figure 10–2. The rule merely sums up all these separate trapezoids. Commencing at the left-hand end of Figure 10–2 the areas of

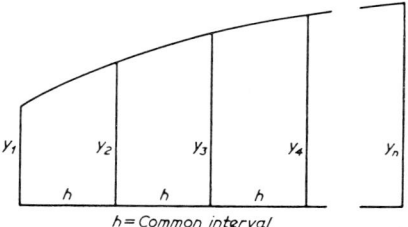

Fig. 10–2—*Trapezoidal rule, common intervals.*

the trapezoids are

$$h/2\,(Y_1 + Y_2);\quad h/2\,(Y_2 + Y_3);\quad h/2\,(Y_3 + Y_4) \text{ etc.}$$

So that the total area of the figure is given by

$$A = h/2\,(Y_1 + 2Y_2 + 2Y_3 + \ldots\ldots + Y_n)$$
$$= h\,(\tfrac{1}{2}Y_1 + Y_2 + Y_3 + \ldots\ldots + \tfrac{1}{2}Y_n)$$

The inside multipliers are only applied to the first and last ordinates and the outside multiplier is unity, so that the rule is simple to use. Thus to determine the area by the trapezoidal rule, divide the figure into a convenient number of equal parts, each of length h. The length of the perpendicular ordinates are measured and substituted in the rule.

AREAS, VOLUMES, MOMENTS, DISPLACEMENT

Obviously the more numerous the ordinates the more accurate will be the answer.

If the ordinates of the figure represent, say, the cross sectional area of a solid then the integration would give the volume of that solid.

It is an advantage to arrange calculations for a particular purpose always in the same form and order and generally tabular. Although in the simple case of Example 10.1 it is not essential to employ the tabular method it nevertheless illustrates the procedure.

EXAMPLE 10.1

The half-breadths at equidistant stations of a ship's waterplane of length 122 metres, commencing from aft, are as follows:

Station   0   1   2   3   4   5   6   7   8   9   10
½ Breadths   2·0 7·3 9·8 10·4 10·6 10·7 10·6 9·9 7·8 4·2 0·2 metres

Determine by the trapezoidal rule the total area of the waterplane.

No. of Ordinate	½ Breadth	Multiplier	Product for Area
0	2·0	½	1·0
1	7·3	1	7·3
2	9·8	1	9·8
3	10·4	1	10·4
4	10·6	1	10·6
5	10·7	1	10·7
6	10·6	1	10·6
7	9·9	1	9·9
8	7·8	1	7·8
9	4·2	1	4·2
10	0·2	½	0·1
			82·4

Common Interval = 122/10 = 12·2 m
½ Area = Common Interval × 82·4 = 12·2 × 82·4 = 1005·3 m².
∴ Total Area = 2 × 1005·3 = 2010·6 m².

Mean Ordinate Rule

The mean ordinate method is a modification of the trapezoidal rule and is frequently used by marine engineers. One half space is used at each end and the common interval is equal to the length (L) of the figure divided by the number of ordinates used. Figure 10–3.

Then
$$A = h(Y_{1\frac{1}{2}} + Y_{2\frac{1}{2}} + Y_{3\frac{1}{2}} + \ldots\ldots Y_{n-\frac{1}{2}})$$

Both external and internal multipliers are equal to unity and thus the rule is simple in use.

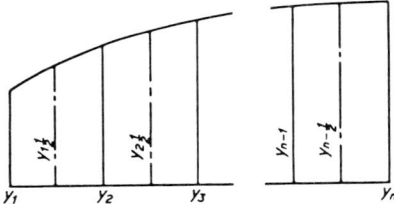

FIG. 10-3—*Mean ordinate rule, common intervals.*

Simpson's Rules

The arithmetical rules for the mensuration of areas, in which one boundary is parabolic, by the use of uniformly spaced ordinates are generally known as Simpson's Rules although they were originally devised by Sir Isaac Newton.

The basis for Simpson's method of deriving the rule associated with a common parabola is shown in Figure 10-4 in which the boundary

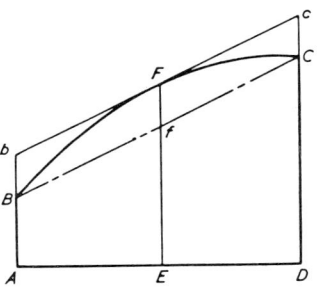

FIG. 10-4—*Simpson's First Rule.*

curve *BFC* is a portion of a parabola. The tangent *bc* is parallel to the chord *BC* and thus makes contact with the parabola at *F*. *FE* is the mid-ordinate. Then

area of circumscribing trapezium $AbcD = EF \times AD = 2EF \times AE$
area of circumscribed trapezium $ABCD = (AB + DC) \times AE$
As the curve *BFC* is a portion of a common parabola
    area $BFCf = 2/3\ AD \times fF =$ say 2 m
Also area $BbFcC = 1/3\ AD \times fF =$ m
then area $ABFCD =$ area $AbcD -$ m
also   = area $ABCD + 2$ m
and $3 \times$ area $ABFCD = (2 \times$ area $AbcD - 2$ m$) +$ (area $ABCD + 2$m)
                    $= (4\ EF \times AE) + [(AB + DC) \times AE]$
                    $= AE\ (4EF + AB + DC)$
with the notation previously used
            $A = h/3\ (Y_1 + 4Y_2 + Y_3)$

This is known as Simpson's First Rule or Three Ordinate Rule. It

AREAS, VOLUMES, MOMENTS, DISPLACEMENT

may be used to determine the area under a curve defined by an *odd* number of equally spaced ordinates. N must be an odd number in order to give an even number of spaces. The multipliers are compounded as shown in Figure 10–5, so that the area under the curve in Figure 10–5 is given by

$A = h/3\ (Y_1 + 4Y_2 + 2Y_3 + 4Y_4 + 2Y_5 + 4Y_6 + 2Y_7 + 4Y_8 + 2Y_9 + 4Y_{10} + Y_{11})$

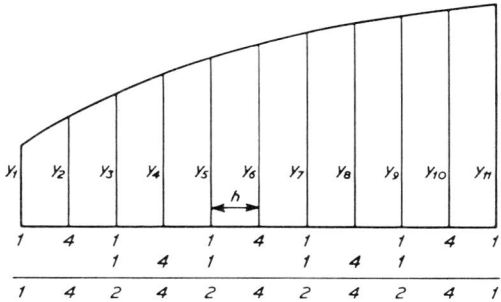

FIG. 10–5—*Simpson's First Rule—compounding of multipliers.*

If the diagram represents one side of a ship's waterplane then the area obtained by the rule will be one-half of the total waterplane area and should therefore be doubled. It is important in all cases to ascertain whether half or whole ordinates are being used.

In the application of the rule it is convenient to arrange the work in tabular form.

EXAMPLE 10.2

The half-breadths of a ship's waterplane at stations 12·2 m apart commencing from aft are 2·0, 7·3, 9·8, 10·4, 10·6, 10·7, 10·6, 9·9, 7·8. 4·2 and 0·2 m respectively. Determine by Simpson's First Rule the total area of the waterplane.

No. of Ordinate	½ Breadth	SM	Product for Area
0	2·0	1	2·0
1	7·3	4	29·2
2	9·8	2	19·6
3	10·4	4	41·6
4	10·6	2	21·2
5	10·7	4	42·8
6	10·6	2	21·2
7	9·9	4	39·6
8	7·8	2	15·6
9	4·2	4	16·8
10	0·2	1	0·2
			249·8

SM = Simpson's Multiplier
Area = 1/3 × 12·2 × 249·8 × 2 for both sides = 2031·7 m²

This result differs from that obtained by the trapezoidal rule by about 1 per cent and should be considered as the more accurate.

Simpson's Second Rule

This rule was derived in the first instance for four evenly spaced ordinates. It assumes the bounding curve to be a parabola of the third order. It may be used to find the area under a curve defined by a number of equally spaced ordinates derived from $(3n + 1)$ where $n = 1, 2, 3$ etc.

For four evenly spaced ordinates the rule is

$$A = \tfrac{3}{8}h\,(Y_1 + 3Y_2 + 3Y_3 + Y_4)$$

A long curvilinear area may be divided into a number of portions similar to the above to each of which the above rule would apply. The multipliers are compounded as shown in Figure 10–6 and the area under the curve in that case is given by

$$A = \tfrac{3}{8}h\,(Y_1 + 3Y_2 + 3Y_3 + 2Y_4 + 3Y_5 + 3Y_6 + 2Y_7 + 3Y_8 + 3Y_9 + Y_{10})$$

EXAMPLE 10.3

The half-ordinates of part of a ship's waterplane are spaced 12 m apart and are 4·2, 7·8, 9·9, 10·5, 10·6, 10·4, 9·8, 7·3, and 2·1 m respectively. Determine the total area by Simpson's 2nd Rule.

No. of Ordinate	½ Ordinate	SM	Product for Area
1	4·2	1	4·2
2	7·8	3	23·4
3	9·9	3	29·7
4	10·5	2	21·0
5	10·6	3	31·8
6	10·6	3	31·8
7	10·4	2	20·8
8	9·8	3	29·4
9	7·3	3	21·9
10	2·1	1	2·1
			216·1

Area = $\tfrac{3}{8}$ × 12 × 216·1 × 2 for both sides
= 1944·9 m²

Simpson's Third Rule or The 5 + 8 − 1 Rule

This is a useful rule which gives the area between any two consecutive ordinates Y_1 and Y_2 when the next equi-spaced ordinate Y_3 is also known. For this rule ordinate Y_1 is known as the "near" ordinate, Y_2 as the "middle" ordinate and Y_3 as the "far" ordinate. In Figure 10–7.

Area $ABCF = h/12\,(5Y_1 + 8Y_2 - Y_3)$ by the "$5 + 8 - 1$" Rule
Also area $CDEF = h/12\,(5Y_3 + 8Y_2 - Y_1)$
Now area $ABDE$ = area $ABCF$ + area $CDEF$
$$= h/12\,(5Y_1 + 8Y_2 - Y_3 + 5Y_3 + 8Y_2 - Y_1)$$
$$= h/12\,(4Y_1 + 16Y_2 + 4Y_3)$$
$$= h/3\,(Y_1 + 4Y_2 + Y_3)$$
which is Simpson's First Rule.

EXAMPLE 10.4

Determine the area between the first pair of ordinates in Figure 10.7 if $Y_1 = 4\cdot 2$ m, $Y_2 = 7\cdot 8$ m and $Y_3 = 9\cdot 9$ m. The common interval $h = 14$ m.

$$\text{Area} = h/12\,(5Y_1 + 8Y_2 - Y_3)$$
$$= 14/12\,[(5 \times 4\cdot 2) + (8 \times 7\cdot 8) - 9\cdot 9]$$
$$= 1\cdot 17\,(21 + 62\cdot 4 - 9\cdot 9)$$
$$= 86 \text{ m}^2$$

Rules can be combined one with another just as the basic unit for each rule is combined in series to deal with many ordinates. Such a rule is the *Six Ordinate Rule* which can be deduced as follows:
From Figure 10–8.

1) Area between Y_1 and $Y_2 = 1/12h[5Y_1 + 8Y_2 - Y_3]$
$$= h[5/12\,Y_1 + 8/12\,Y_2 - 1/12\,Y_3]$$
2) Area between Y_2 and $Y_5 = 3/8h[Y_2 + 3Y_3 + 3Y_4 + Y_5]$
$$= h[3/8\,Y_2 + 9/8\,Y_3 + 9/8\,Y_4 + 3/8\,Y_5]$$

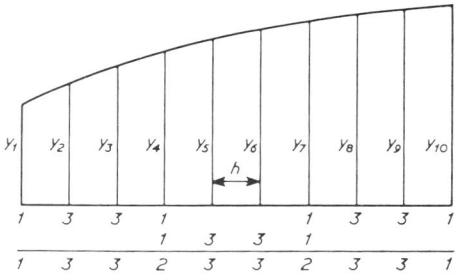

FIG. 10–6—*Simpson's Second Rule—compounding of multipliers.*

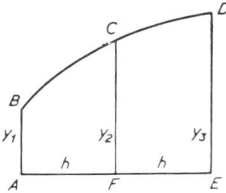

FIG. 10–7—*Simpson's Third Rule.*

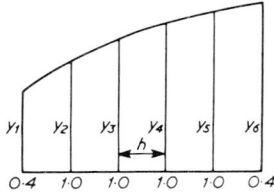

FIG. 10-8—*Six Ordinate Rule.*

3) Area between Y_5 and $Y_6 = 1/12h[5Y_6 + 8Y_5 - Y_4]$
$= h[5/12 Y_6 + 8/12 Y_5 - 1/12 Y_4]$

Total area $= [1] + [2] + [3]$
$= h[5/12 Y_1 + 25/24 Y_2 + 25/24 Y_3 + 25/24 Y_4 + 25/24 Y_5 + 5/12 Y_6]$
$= \dfrac{25}{24} h[0\cdot4 Y_1 + Y_2 + Y_3 + Y_4 + Y_5 + 0\cdot4 Y_8]$

There are instances where neither the First nor the Second of Simpson's Rules fit the circumstances, such as for example where there are eight equi-spaced ordinates. This can be resolved by dealing with the area in two parts

1) Determine the area between Y_1 and Y_5 using the First Rule.
2) Determine the area between Y_5 and Y_8 using the Second Rule.

The total area = sum of areas from (1) and (2)

Intermediate Ordinates

Towards the end of a waterline the curvature is sometimes so great that a closer spacing of ordinates is necessary to give a sufficiently high standard of accuracy. If fractional spacing is to be used in conjunction with normal spacing the appropriate internal multipliers are divided by the required fraction. If the common interval is halved the internal multipliers are also halved over that part of the range. In this way the calculation can be kept continuous. The internal multipliers, where the normal spacing is halved, are compounded as shown in Figure 10–9.

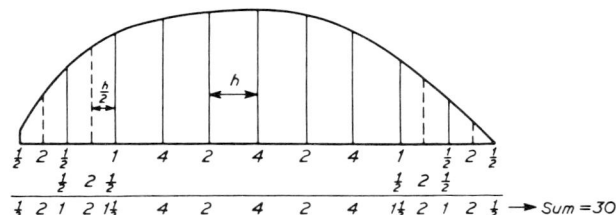

FIG. 10-9—*Compounding of internal multipliers.*

A useful check on the accuracy of an amended rule is that the sum

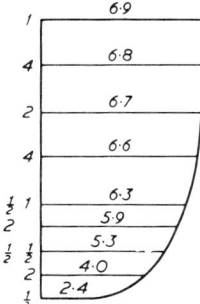

FIG. 10–10—*Half ordinates of ship's bulkhead (Example 10.5.)*

of all the internal multipliers must equal the number of whole spaces divided by the external multiplier. In Figure 10–9 the sum of internal multipliers = 30. Number of whole spaces divided by external multiplier = 10 ÷ 1/3rd = 30.

It should be noted that half-spacing is only suitable for either Simpson's First Rule or the Trapezoidal methods.

It is assumed that the area to be measured is bounded by a continuous curve. If there is a break or "knuckle" on the curve a dividing ordinate should be placed there and the other ordinates arranged to suit.

EXAMPLE 10.5

The half-ordinates of a bulkhead in a ship Figure 10–10, commencing at the top, are 6·9, 6·8, 6·7, 6·6, 6·2, 5·9, 5·3. 4·0 and 2·4 m in length respectively. The common interval between ordinates is 1 m between the first and fifth ordinates and ½ m between the fifth and ninth. Determine the area of the bulkhead.

No. of Ordinates	½ Ordinates	SM	Products for Area
1	6·9	1	6·9
2	6·8	4	27·2
3	6·7	2	13·4
4	6·6	4	26·4
5	6·2	1½	9·3
6	5·9	2	11·8
7	5·3	1	5·3
8	4·0	2	8·0
9	2·4	½	1·2
		18	109·5

Figure 10–10. Sum of internal multipliers = 18. Number of whole spaces = 6 and 6 ÷ ½ = 18
Total area = 1/3 × 1 × 109·5 × 2 for both sides = 73 m^2

Tchebycheff's Rules

This system differs from the Simpson rules in that the ordinates are not equi-spared. The boundary curve is assumed to be parabolic and the spacing is such that the ordinates do not require internal multipliers but can be directly added. The spacing of the ordinates is such that the average of the measured ordinates is the mean ordinate of the figure. The end ordinates of the figure are not used. The area is then obtained by adding together the lengths of the measured ordinates, dividing by the number of ordinates and finally multiplying by the length of the figure for which the area is desired.

With an even number of measured ordinates there is no middle ordinate at the centre of the base.

The spacing to be used is given in Table 10.2 in fractions of the half-length of base and is measured from the middle of the base.

Table 10.2—Tchebycheff's Rules

Length of base $= L$; No. of ordinates used $= N$; Area $= L/N$ (sum of measured ordinates)

No. of ordinates used	Position of ordinates from the middle of the base in fractions of the half length					
N	amidships	1	2	3	4	5
2		0·5773				
3	0	0·7071				
4		0·1876	0·7947			
5	0	0·3745	0·8325			
6		0·2666	0·4225	0·8662		
7	0	0·3239	0·5297	0·8839		
8		0·1026	0·4062	0·5938	0·8974	
9	0	0·1679	0·5288	0·6010	0·9116	
10		0·0838	0·3127	0·5000	0·6873	0·9162

The position of the ordinates in the case of the four ordinate rule is shown in Figure 10–11. For the four ordinate rule.

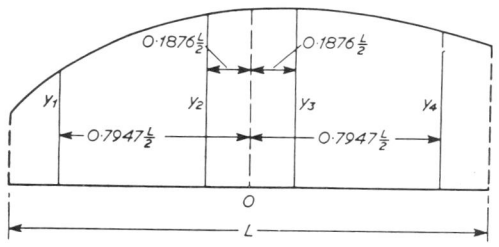

Fig. 10–11—*Tchebycheff's Four Ordinate Rule.*

$$\text{Area} = \frac{Y_1 + Y_2 + Y_3 + Y_4}{4} \times L$$

EXAMPLE 10.6

For the part of a waterplane shown in Figure 10-11 the half-ordinates for Tchebycheff's four ordinate rule are as follows: $Y_1 = 3\cdot1$ m; $Y_2 = 4\cdot6$ m; $Y_3 = 4\cdot7$ m; $Y_4 = 4\cdot1$ m. The length L is 25 m. Determine the total area for the part waterplane.

$$\text{Total area} = \frac{3\cdot1 + 4\cdot6 + 4\cdot7 + 4\cdot1}{4} \times 25 \times 2 \text{ for both sides}$$

$$= 206\cdot25 \text{ m}^2$$

Mechanical Integrators

The foregoing has been concerned with the method of performing integration by the use of arithmetical rules. When the area to be determined has a bounding curve that embodies discontinuities at irregular intervals the process of arithmetical integration becomes laborious. To simplify the measurement of the properties of plane figures several types of mechanical integrator have been developed, of which three are in fairly common use in naval architecture; they are the planimeter, integrator and integraph.

The Planimeter

The planimeter is the simplest form of mechanical integrator. In a single operation it can measure the area enclosed by any curve. It consists of two arms jointed or hinged together. One arm, the pole arm, is a radius turning about a needle-pointed weight. The other arm, the tracer arm, has a pointer at its free end and wheels at the other end from which the area is determined. The wheel records only the lateral displacements of the tracer arm, taking no account of lengthwise movements.

To measure an area the needle point is pressed into the paper, generally outside the boundary of the figure to be measured. The tracing point is placed at some convenient point on the boundary of the figure and the wheel reading noted. The tracing point is then guided around the outline of the figure and on return to the starting point the wheel reading is again noted. The difference between the final and initial readings of the wheels when multiplied by a simple factor gives the area.

One form of the planimeter is shown diagrammatically in Figure 10-12. A sliding bar pattern as manufactured by W. F. Stanley & Co. Ltd. of London is illustrated in Figure 10-13.

The Integrator

The integrator measures in addition to areas, the first and second moments of the areas about a chosen datum line. It comprises a frame-

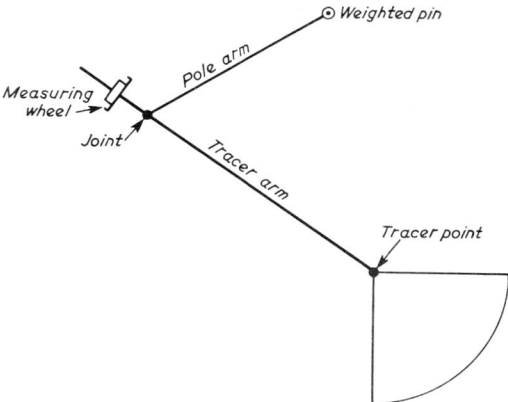

FIG. 10–12—*Planimeter.*

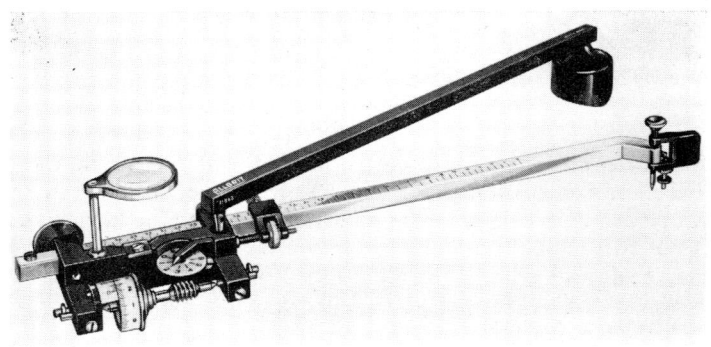

FIG. 10–13—*Stanley Planimeter.*

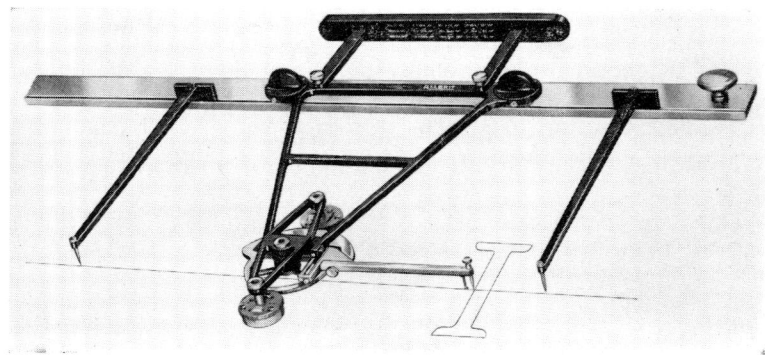

FIG. 10–14—*Stanley Planimeter.*

work constrained to move along a grooved bar *BB* which is set at the required distance from the chosen axis *XX* by means of distance pieces *YY*. The Stanley integrator and a diagrammatic representation are shown in Figures 10–14 and 10–15 respectively. After the tracing point *P*

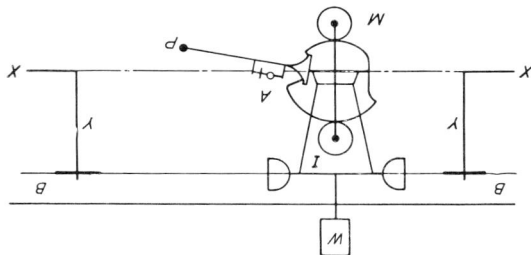

FIG. 10–15—*Stanley Integrator.*

has been guided around the boundary of an area the wheel *A* registers the area, wheel *M* registers the moment of area about axis *XX* and wheel *I* in conjunction with the area wheels registers the moment of inertia about *XX*.

The Integraph

The integraph performs integration graphically. It draws the "first integral curve" of any curve. The Stanley integraph is shown in Figure 10–16. If the tracing point of the integraph is taken along a curve such

FIG. 10–16—*Stanley Integraph.*

as *ABC* in Figure 10–17 a pen on the machine draws in the first integral curve as *abc*. So that at any point *N* on curve *ABC* having an ordinate *Y* the corresponding point *n* on the new curve of ordinate Y_1, will give to some scale the shaded area under the original curve up to point *N*.

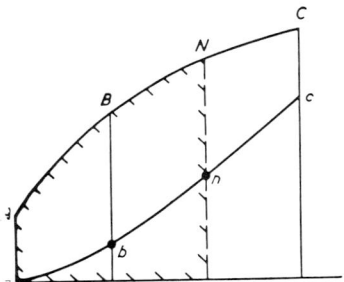

FIG. 10-17—*Curve drawn by Integraph.*

Moments and Centroids

The calculations in naval architecture require the computing of moments of area about chosen axes and the determination of the positions of centroids in relation to these axes; it is also necessary to determine the second moments of the areas or the moments of inertia about axes through the centroid.

Associated with the rules for areas are rules for moments of areas and hence the position of the centroid or centre of area by

Distance of centroid from a given axis =

$$\frac{\textit{Moment of area about the Axis}}{\textit{Area}}$$

The moments of areas are required about two principal axes:
1) one which is transverse and
2) one which is longitudinal

1) *Centroid* of a curvilinear area related to one of its ordinates.

In Figure 10-18 the area of an elemental strip is $y\,dx$ and the moment of the strip about ordinate OA is $y\,x\,dx$.

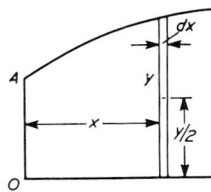

FIG. 10-18—*Illustration of elemental strip.*

$$\text{Moment of total area about } OA = \int y\,x\,dx$$

and

$$\text{Total area} = \int y\,dx.$$

AREAS, VOLUMES, MOMENTS, DISPLACEMENT   141

2) *Centroid* of area related to base. Figure 10–18.
On the assumption that the elemental strip is a rectangle, then its centroid is $y/2$ from the base and the moment of the strip about the base is $\frac{1}{2}y^2 dx$

Moment ot Total area about base $= \int \frac{1}{2}y^2 dx$

and   Total area $= \int y\,dx$

Moment of Area and Position of Centroid by the Use of Simpson's First Rule

1) The example given below is for a ship's waterplane. It is convenient, to express the position of the centroid with respect to the midlength, i.e. the midship station. The levers are measured from this position and for arithmetical simplicity the number of intervals is adopted and the final result multiplied by the common interval. If the product for moments mA is greater than mF then the centroid will be abaft the midship station. If mF is greater than mA then the reverse will be the case.

EXAMPLE 10.7

The half-breadths in metres of a waterplane at stations 12·2 metres apart are 2·0, 7·3, 9·8, 10·4, 10·6, 10·7, 10·6, 9·9, 7·8, 4·2 and 0·2 respectively commencing at the aft end. Determine the position of the centroid relative to the midship (⊛) station.

	Ordinate	½ Breadth	SM	Products for area	Lever from amidships	Products for moments
Aft	0	2·0	1	2·0	5	10·0
	1	7·3	4	29·2	4	116·8
	2	9·8	2	19·6	3	58·8
	3	10·4	4	41·6	2	83·2
	4	10·6	2	21·2	1	21·2
⊛	5	10·7	4	42·8	0	$\overline{290·0} = mA$
	6	10·6	2	21·2	1	21·2
	7	9·9	4	39·6	2	79·2
	8	7·8	2	15·6	3	46·8
	9	4·2	4	16·8	4	67·2
Forward	10	0·2	1	0·2	5	1·0
				249·8		215·4 = mF
						290·0 = mA
					Excess =	74·6 aft

Area of waterplane = 1/3 × 12·2 × 249·8 × 2 for both sides = 2031·7 m²
Moment of area about amidships = 1/3 × 12·2 × 12·2 × 74·6 × 2 for both sides

$$= 7402 \cdot 3 \text{ m}^2\text{m}$$

Centroid from amidships $= \dfrac{7402 \cdot 3}{2031 \cdot 7} = 3 \cdot 64$ m abaft amidships

or due to cancellations

Centroid from amidships $= \dfrac{74 \cdot 6}{249 \cdot 8} \times 12 \cdot 2$

$= 3 \cdot 64$ m abaft amidships

NOTE:
The centroid of a waterplane is known as the *Centre of Flotation*

2) To determine the position of the centroid of area *Relative to base*.

EXAMPLE 10.8
The ordinates of a rudder abaft the axis at intervals of 0·75 m commencing from the top are 0·76, 3·05, 3·70, 3·80, 3·65, 2·90 and 0·60 m respectively. Determine the area of the rudder and the position of the centroid relative to the axis.

The quantity required is $\overline{Y}$ as indicated in Figure 10–19.

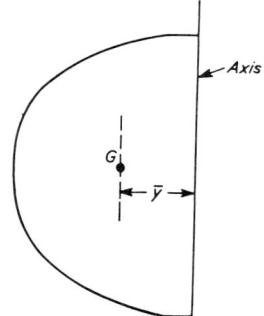

FIG. 10–19—*See example 10.8.*

Ordinate	SM	Products for area	(Ord)²	SM	Products for Moment about Axis
0·76	1	0·76	0·6	1	0·6
3·05	4	12·20	9·3	4	37·2
3·70	2	7·40	13·7	2	27·4
3·80	4	15·20	14·4	4	57·6
3·65	2	7·30	13·3	2	26·6
2·90	4	11·60	8·4	4	33·6
0·60	1	0·60	0·4	1	0·4
		55·06			183·4

Area = 1/3 × 0·75 × 55·06 = 13·76 m²
Moment of area about axis = $\frac{1}{2} \times \frac{1}{3} \times 0.75 \times 183.4 = 22.93$
Centroid from Axis = $\bar{Y} = \dfrac{22 \cdot 93}{13 \cdot 76} = 1 \cdot 67$ m

3) *Three—Ten—Minus—One Moment Rule*

The longitudinal moment of an area between two consecutive ordinates say Y_1 and Y_2 is sometimes required. If an adjacent ordinate Y_3 is known the moment of area between Y_1 and Y_2 about Y_1 is given by

$$Moment = 1/24\ h^2\ (3Y_1 + 10Y_2 - Y_3)$$

EXAMPLE 10.9

A curve has ordinates $Y_1 = 3 \cdot 05$ m; $Y_2 = 2 \cdot 75$ m; $Y_3 = 2 \cdot 15$ m. The common interval = 1·2 m. Determine the position of the centroid of the portion between Y_1 and Y_2 with respect to Y_1

Ordinate		Area		Moment	
		SM	Products	SM	Products
Y_1	3·05	5	15·25	3	9·15
Y_2	2·75	8	22·0	10	27·50
Y_3	2·15	−1	−2·15	−1	−2·15
			35·10		34·50

Area = 1/12 × 1·2 × 35·10 = 3·51 m²
Moment = 1/24 × 1·2² × 34·50 = 2·07 m²m
Centroid from $Y_1 = \dfrac{2 \cdot 07}{3 \cdot 51} = 0 \cdot 59$ m

Moment of Inertia

The second moment of an area, generally known as the moment of inertia (I) of an area about an axis is equal to the sum of all the elements of the area times the square of the distance of the element from the axis and is given by

$$I \text{ about } oy = \int x^2 y\,dx$$

$$I \text{ about } ox = \tfrac{1}{3} \int y^3\,dx$$

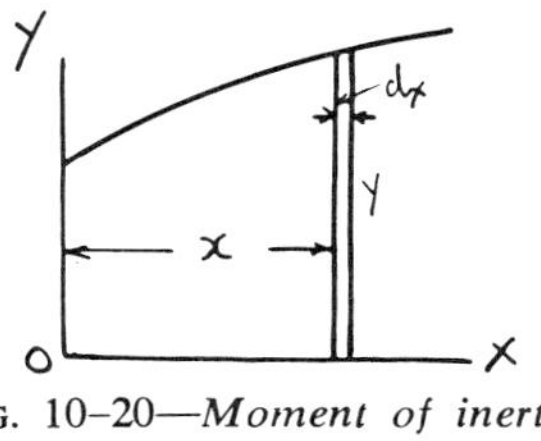

FIG. 10–20—*Moment of inertia.*

The I of an area about an axis (OO) parallel to and a given distance h from an axis NA passing through the centroid of the area is obtained in the following manner.

Let I_G = moment of inertia of area about axis NA passing through centroid
h = distance of given axis from NA
Then I of area (A) about axis OO is given by

$$I = I_G + Ah^2 \quad \text{[Theorem of Parallel Axes]}$$

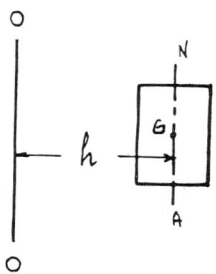

FIG. 10-21—*Theorem of parallel axes.*

Thus the moment of inertia of an area about an axis through its own centroid is always less than that about any other axis parallel to it.

The moment of inertia of some regular areas can be expressed in simple formulae as follows:

1) *Rectangle*

The moment of inertia of a rectangle about an axis in its plane, parallel to one side, and passing through its centroid [C] is

$$I_{xx} = \frac{BD^3}{12}$$

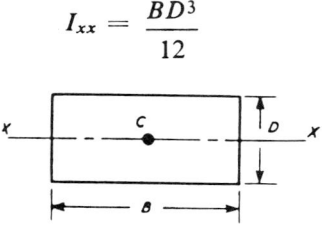

FIG. 10-22—*Moment of inertia of a rectangle.*

If the axis coincides with one side

$$I_{xx} = \frac{BD^3}{3}$$

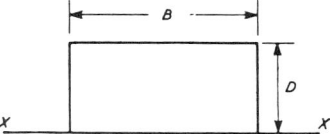

FIG. 10–23—*Moment of inertia of a rectangle, axis coincident with one side.*

2) Triangle

The moment of inertia of a triangle about an axis in its plane parallel to one side and passing through its centroid is

$$I_{xx} = \frac{BH^3}{36}$$

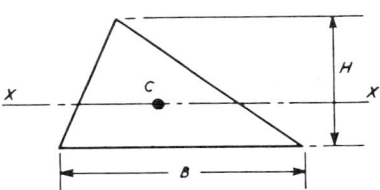

FIG. 10–24—*Moment of inertia of a triangle.*

If the axis coincides with one side

$$I_{xx} = \frac{BH^3}{6}$$

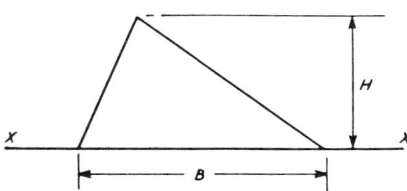

FIG. 10–25—*Moment of inertia of a triangle, axis coincident with one side.*

3) Circle

The moment of inertia of a circle of diameter D about an axis in its plane and passing through the centre is

$$I_{xx} = \frac{\pi D^4}{64}$$

146 SHIPS AND NAVAL ARCHITECTURE

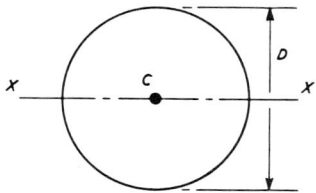

FIG. 10-26—*Moment of inertia of a circle.*

If the axis is perpendicular to the plane of the circle and passes through its centre the moment of inertia is called the polar moment of inertia [*J*] and is

$$J = \frac{\pi D^4}{32}$$

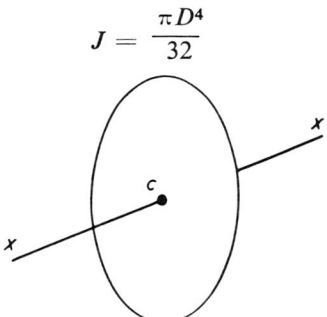

FIG. 10-27—*Polar moment of inertia (J).*

The Moment of Inertia of a Ship's Waterplane about the Fore and Aft Centre Line. (Transverse Moment of Inertia).

The calculation for the transverse moment of inertia may be made as in the following example.

EXAMPLE 10.10

The half-ordinates of a ship's waterplane are 6 metres apart and are 0, 2·1, 3·8, 5·2, 5·1, 4·7, 3·5, 1·9 and 0 m respectively. Determine the transverse moment of inertia of the waterplane about the centre-line.

½ Ord Y	[½ Ord]³ Y³	SM	Products for MI
0	0	1	—
2·1	9	4	36
3·8	55	2	110
5·2	141	4	564
5·1	133	2	266
4·7	104	4	416
3·5	43	2	86
1·9	7	4	28
0	0	1	—
			1506

AREAS, VOLUMES, MOMENTS, DISPLACEMENT   147

$$I = \tfrac{1}{3}\int Y^3 dx$$
$$= \tfrac{1}{3} \times 1506 \times \tfrac{1}{3} \times 6 \times 2$$
$$= 2008 \text{ m}^4$$

$\tfrac{1}{3}$ for SM;   6 for common interval $= dx$;   2 for both sides.

The moment of inertia of waterplane about the centre of flotation $[I_{CF}]$
By the Theorem of Parallel Axes given above
$I_{\text{amidship}} = I_{CF} + Ah^2$

or

$I_{CF} = I_{\text{amidship}} - Ah^2$

where

A = area of waterplane
h = distance of centre of flotation [CF] from amidships [⊗]

EXAMPLE 10.11.

The half-breadths of a ship's waterplane at stations 10·5 m apart and commencing from aft are as follows:

Station	0 [Aft]	1	2	3	4	5	6	7	8	9	10 [Forward]	
½ Breadths	0·2	7·4	8·7	9·0	9·1	9·2	9·1	8·6	7·8	5·1	0	metres

Determine 1) the area of the waterplane

2) the position of the centre of flotation relative to amidships

and   3) the moment of inertia about the centre of flotation

Station	½ breadths	SM	Products for area	Lever	Products for moments	Lever	Products for moments of inertia
0[aft]	0·2	1	0·2	5	1·0	5	5·0
1	7·4	4	29·6	4	118·4	4	473·6
2	8·7	2	17·4	3	52·2	3	156·6
3	9·0	4	36·0	2	72·0	2	144·0
4	9·1	2	18·2	1	18·2	1	18·2
⊗5	9·2	4	36·8	0	$\overline{261·8}$ = mA	0	0
6	9·1	2	18·2	1	18·2	1	18·2
7	8·6	4	34·4	2	68·8	2	137·6
8	7·8	2	15·6	3	46·8	3	140·4
9	5·1	4	20·4	4	81·6	4	326·4
10 [Forward]	0	1	—	5	—	5	—
			$\overline{226·8}$ = S		$\overline{215·4}$ = mF		$\overline{1420·0}$ = Z
					$261·8$ = mA		
				Excess	46·4 mA		

1) Area $= \tfrac{1}{3} \times 10\cdot5 \times 226\cdot8 \times 2$ for both sides
   $= 1588$ m² $= A$

2) CF from amidships $=$
   $$\frac{[mA \sim mF] \times CI}{S} = \frac{46\cdot4}{226\cdot8} \times 10\cdot5 = 2\cdot14 \text{ m abaft } ⊗ = h$$

3) $I_{\text{amidships}} = \tfrac{1}{3}[CI]^3 \times Z \times 2$ for both sides
   $= \tfrac{1}{3} \times 10\cdot5^3 \times 1420 \times 2 = 1{,}096{,}000$ m⁴
   $I_{CF} = I_{\text{amidships}} - Ah^2 = 1{,}096{,}000 - 1588 \times 2\cdot14^2$
   $= 1{,}088{,}700$ m⁴

Volumes and Centroids of Volumes

The determination of volumes and positions of their centroids is simply an application of the method of calculating an area and the position of the centroid of the area.

The distribution of the volume of a ship can be represented by a curve, any ordinate of the curve representing, to scale, the area of immersed section of the ship at the corresponding position in the ship's length.

Figure 10–28 represents the "sectional area" curve for a ship to the

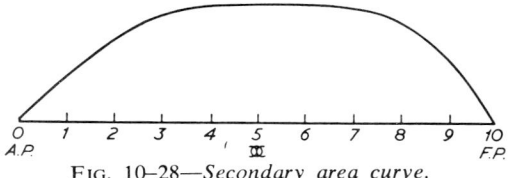

FIG. 10–28—*Secondary area curve.*

load waterline. The area under the curve is a measure of the volume of the ship to the load waterline and the longitudinal position of the centroid of the sectional area curve is the same as the longitudinal centroid of the volume.

Figure 10–29 represents one side of part of the after body of a ship for which the volume and position of centroid are desired.

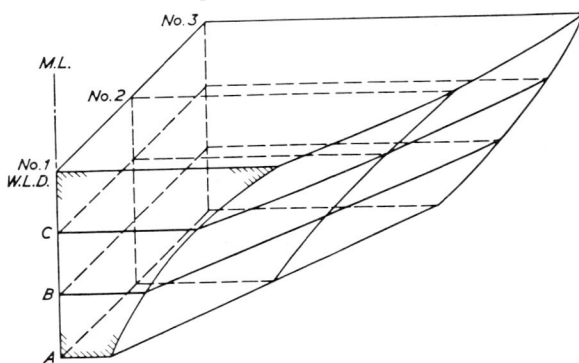

FIG. 10–29—*Determination of centroid, one side of after body given.*

AREAS, VOLUMES, MOMENTS, DISPLACEMENT

The volume can be determined by
1) measuring the areas of the vertical stations 1, 2 and 3 or
2) measuring the areas of the horizontal planes A, B, C and D.

Then one or other of the groups of areas is integrated using one of the arithmetical rules—Simpson's—already described.

The vertical stations are 2·13 m apart and the waterlines are spaced 0·61 m.

EXAMPLE 10.12

The areas of the waterplanes and transverse sections for one side of Figure 10.29 are as follows:

WL	Area (m²)	Section	Area (m²)
A	5·8	1	2·1
B	8·0	2	4·1
C	10·7	3	5·7
D	13·3		

Determine the total volume and the position of the centroid both vertically relative to WL A and longitudinally relative to Section No. 1.

1)	Section	Area (m)²	SM	Products	Lever	Products
	1	2·1	1	2·1	0	—
	2	4·1	4	16·4	1	16·4
	3	5·7	1	5·7	2	11·4
				24·2		27·8

Volume = ⅓ × 2·13 × 24·2 × 2 for both sides
= 34·4 m³

Centroid from Section No. 1 = $\frac{27·8}{24·2}$ × 2·13 = 2·45 m

2)	WL	Area (m)²	SM	Products	Lever	Products
	A	5·8	1	5·8	0	—
	B	8·0	3	24·0	1	24·0
	C	10·7	3	32·1	2	64·2
	D	13·3	1	13·3	3	39·9
				75·2		128·1

Volume = ⅜ × 0·61 × 75·2 × 2 for both sides
= 34·4 m³ as in [1]

Centroid above WL A = $\frac{128·1}{75·2}$ × 0·61
= 1·04 m

Since the ship is symmetrical about the middle line plane the centroid is in the middle line plane 2·45 m from Section No. 1 and 1·04 m above WL A.

Displacement

The volume of displacement is the total volume of fluid displaced by the ship. This can be readily appreciated by imagining the fluid to be wax and the ship removed from it; it is then the volume of the impressions left by the hull.

The Archimedes principle states that when a solid is immersed in a liquid it experiences an upthrust equal to the weight of the fluid displaced. This upthrust is called the buoyancy of the solid. A corollary of the Archimedes principle, known as the *Law of Flotation*, states that when a body is floating freely in a fluid the weight of the body equals the buoyancy, which is the weight of the fluid displaced.

In the case of a floating ship this upward force or buoyancy is most commonly called the displacement and is equal to the volume of displacement multiplied by the density of the fluid. The displacement is denoted by the symbol Δ and the volume of displacement by the symbol ∇. The letter V is used to denote speed. Since the displacement must equal the weight the terms are frequently used synonymously in the form of light displacement, ballast displacement, deep load displacement etc. These terms are being replaced by light mass, ballast mass, deep mass etc.

The forces acting on a floating body are
1) The weight vertically downwards which can be taken as acting at the centre of gravity.
2) The buoyancy vertically upwards which can be taken as acting at the centre of buoyancy which is the centroid of the immersed volume.

Displacement is a force and should, therefore, in SI units, be expressed in newtons or meganewtons. However, it has been accepted that for the convenience of ship operators and others, the metric tonne force be used for displacement.

The mass density of freshwater (FW) is 1,000 kg/m^3
The mass density of sea water (SW) is 1,025 kg/m^3
and since 1,000 kg = 1 tonne

Displacement in Fresh Water in Tonnes = immersed volume (m^3) × 1·000
Displacement in Sea Water in Tonnes = immersed volume (m^3) × 1·025

For some purposes it is necessary to determine the precise position of the centroid of the immersed volume i.e. the *Centre of Buoyancy*. This involves the determination of the vertical position, generally relative to the base line or keel (KB) and the longitudinal position in the middle line plane relative to amidships (LCB).

The procedure follows that of Example 10.12 and in more detail as follows:

AREAS, VOLUMES, MOMENTS, DISPLACEMENT 151

EXAMPLE 10.13

A ship of length 130 m floats at a uniform draught of 6·5 m. The areas of waterplanes at 1·3 m apart and areas of immersed sections at equi-spaced stations are as follows:

Waterplane   1·3   2·6   3·9   5·2   6·5   metres above base line
Areas   1460   1630   1740   1790   1800   m²
Station   0 (aft) 1   2   3   4   5   6   7   8   9   10 (forward)
Areas   —   37   78   100·5   107   107·8   107·5   105   87   44·2   —   m²

The appendage below 1·3 m waterline has a displacement of 1310 tonnes with a KB of 0·67 m.

Determine 1) the displacement of the ship in sea water
           2) the position of the centre of buoyancy above the keel (KB) and position relative to amidships (LCB)

Waterplane Metres	Area m²	SM	Products of Areas	Levers about Base	Moments of Products of Areas
1·3	1460	1	1460	1	1460
2·6	1630	4	6520	2	13040
3·9	1740	2	3480	3	10440
5·2	1790	4	7160	4	28640
6·5	1800	1	1800	5	9000
			20420		62580

Displacement of main hull in sea water between 1·3 and 6·5 m waterplanes
= ⅓ × 1·3 × 20420 × 1·025 = 9070 tonnes

KB of main hull above base = $\frac{62580}{20420} \times 1·3 = 3·98$ m

Item	Displ. [Tonnes]	KB [m]	Moment [Tonne m]
Main Hull	9070	3·98	36100
Appendage	1310	0·67	878
Total Disp. =	10380		36978

Centre of Buoyancy above base [KB] = $\frac{36978}{10380} = 3·56$ m

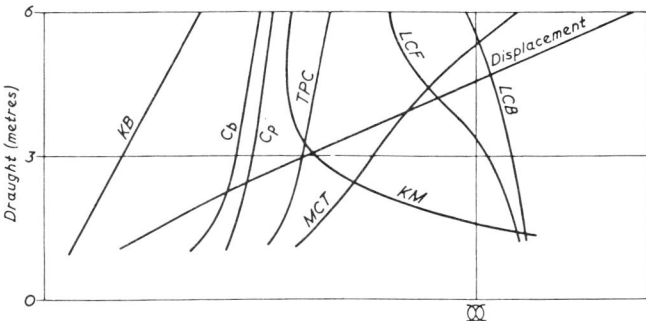

FIG. 10-30—*Hydrostatic curves.*

	Station	Area [m²]	SM	Products of Areas	Levers about ⊕	Moments of Products of Areas
Aft	0	0	1	—	5	—
	1	37	4	148	4	592
	2	78	2	156	3	468
	3	100·5	4	402	2	804
	4	107	2	214	1	214
⊕	5	107·8	4	431·2	0	2078 mA
	6	107·5	2	215	1	215
	7	105	4	420	2	840
	8	87	2	174	3	522
	9	44·2	4	176·8	4	707·2
Forward	10	—	1	—	5	—
				2337·0		2284·2 mF
						2078·0 mA
					Excess =	206·2 mF

Total displacement in Sea water = ⅓ × 130/10 × 2337 × 1·025 = 10380
Tonnes as above

Longitudinal position of centre of buoyancy =

$$\frac{206·2}{2337} \times \frac{130}{10} = 1·15 \text{ m forward of amidships.}$$

The calculations for displacement, centres of buoyancy and flotation etc. are carried out to a series of equally spaced waterlines. These calculations are commonly made on a *Displacement Sheet*. Collectively the information derived is known as the hydrostatic data and is presented either in tabular form or as a set of curves called the *Hydrostatic Curves*.

The traditional manner of calculating the hydrostatic data is lengthy and laborious. However, it is so methodical that the process of direct calculation has been replaced by computer processing.

Typical hydrostatic curves are shown in Figure 10–30.

Centre of Buoyancy from Displacement Curve
The area of a curve of displacement up to a given waterline divided by the displacement at that waterline gives the distance of the centre of buoyancy below the waterline.

EXAMPLE 10.14
The displacements of a ship at waterlines 0·65 m apart are 1330, 1050, 770, 520, 285, 90 and 0 tonnes. Determine the distance of the centre of buoyancy above the keel [KB] at the 3·90 m waterline.

AREAS, VOLUMES, MOMENTS, DISPLACEMENT 153

WL	Δ	SM	Functions
Base	—	1	—
0·65	90	4	360
1·30	285	2	570
1·95	520	4	2080
2·60	770	2	1540
3·25	1050	4	4200
3·90	1330	1	1330
			10080

Area under displacement curve $= \frac{1}{3} \times 0.65 \times 10080 = 2184$

Distance of centre of buoyancy below 3·90 m waterline $= \frac{2184}{1330} = 1.64$ m

$$\text{Draught} = 3.90$$
$$\text{KB} = \overline{2.26} \text{ m}$$

Tonnes per Centimetre (TPC) is the mass to be added or deducted from a ship to change the mean draught by 1 cm.

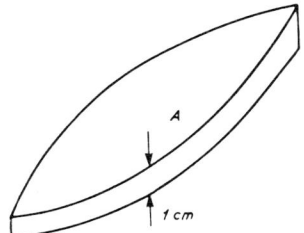

FIG. 10–31—*Tonnes per centimetre immersion.*

Let A = area of a waterplane in m²
Assume a layer 1 cm thick, then the volume of this element equals

$$A \times \frac{1}{100} \text{ m}^3$$

the displacement of this layer in sea water

$$= A \times \frac{1}{100} \times 1.025$$

or $\quad = A \times \frac{1}{100} \times \frac{1}{0.975} = \frac{A}{97.5} =$ tonnes per cm

in fresh water $= A \times \frac{1}{100} \times 1.000 = \frac{A}{100} =$ tonnes per cm

Thus for an additional mass w the sinkage is given approximately by

w/TPC; for wall-sided vessels this expression is precise. This is a very useful expression and the values for TPC are calculated for all waterplanes and they form an important item in the hydrostatic data.

If the mass is placed vertically above or below the centre of flotation of the waterplane at which the vessel floats then the ship will sink to a new waterline parallel to the original waterline. Placed at any other position the ship will change trim as well as sink bodily in the water. This is dealt with in a later chapter.

EXAMPLE 10.15
If the TPC for a ship at a particular waterline is 20·8 determine:
a) the mass required to increase the mean draught by 15 cm.
b) the increase in mean draught if 550 tonnes are placed on board.
a) Mass required = 15 × 20·8 = 312 tonnes.
b) Increase in mean draught = $\dfrac{550}{20 \cdot 8}$ = 26·4 cm

The *Displacement* and *KB* can be determined from the TPC values as indicated in Example 10.16.

$$\text{Since TPC} = \frac{\text{Area of waterplane}}{97 \cdot 5}$$

then the waterplane areas are given by TPC × 97·5 and these can be used to determine the displacement as already shown in Example 10.13. To reduce the arithmetical work the multiplication by 97·5 is introduced at the end.

EXAMPLE 10.16
The TPC values for a ship at waterplanes 0·9 m apart, from the keel to the load waterplane are as follows:
0, 7·6, 10·0, 11·6, 12·6, 13·4, and 14·0.
Determine the displacement and the *KB* at the load draught.

	TPC	SM	Products	Lever	Products
Keel	0	1	—	0	—
	7·6	4	30·4	1	30·4
	10·0	2	20·0	2	40·0
	11·6	4	46·4	3	139·2
	12·6	2	25·2	4	100·8
	13·4	4	53·6	5	268·0
	14·0	1	14·0	6	84·0
			189·6		662·4

Displacement to load draught of 5·4 m =
$$1/3 \times 0 \cdot 9 \times 189 \cdot 6 \times 97 \cdot 5 \times 1 \cdot 025 = 5694 \text{ tonnes.}$$

KB at the load draught = $\dfrac{662 \cdot 4}{189 \cdot 6} \times 0 \cdot 9 = 3 \cdot 14$ m

Form Coefficients

Form is used in general terms to describe the shape of the hull of a ship and compare one ship with another. It is expressed as coefficients such as: *The Block Coefficient* (C_b) is the ratio of the volume of displacement to a given waterline and the volume of the circumscribing block of constant rectangular section having the same length, breadth and draught as the ship.

$$C_b = \frac{\overline{V}}{LBH}$$

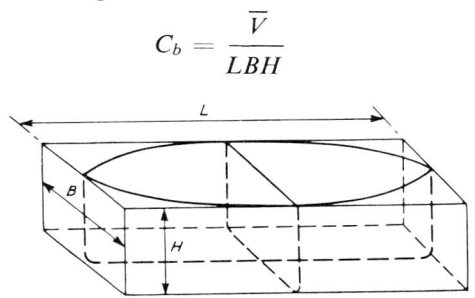

L = Length of ship between perpendiculars (L.B.P.) in metres
B = Breadth moulded in metres
H = Mean draught moulded in metres

FIG. 10-32—*Block coefficient.*

From this it follows that if dimensions are in metres then

Displacement in tonnes (SW) $= L \times B \times H \times C_b \times 1 \cdot 025$

The Prismatic Coefficient (C_p) is the ratio of the volume of displacement of the ship to the volume of the circumscribing block having a constant section equal to the immersed midship section of the ship (A_m) and the length (L) of the ship.

$$C_p = \frac{\overline{V}}{LA_m}$$

The Midship Section Area Coefficient (C_m) is the ratio of the immersed area of the midship section to the area of the circumscribing rectangle having a breadth equal to that of the ship and depth equal to the draught

$$C_m = \frac{A_m}{B \times H}$$

There is a relation connecting the block, prismatic and midship area coefficients

$$C_p = \frac{\overline{V}}{LA_m} = \frac{\overline{V}}{LBH \, C_m} = \frac{C_b}{C_m}$$

In other words
$$C_b = C_p \times C_m$$

Waterplane Area Coefficient (C_w) is the ratio of the area of the waterplane (A_w) to the area of the circumscribing rectangle having a length equal to that of the ship and a breadth equal to that of the ship.

$$C_w = \frac{A_w}{L \times B}$$

Vertical Prismatic Coefficient (C_{pv}) is the ratio of the immersed volume to the area of the load waterplane multiplied by the mean draught.

$$C_{pv} = \frac{\overline{V}}{A_w \times H}$$

or
$$= \frac{C_b}{C_w}$$

The vertical prismatic indicates the vertical distribution of the displacement.

A Mean Ordinate is defined by the following equation:

$$\text{The mean ordinate of a figure} = \frac{\text{area of the figure}}{\text{length of base}}$$

EXAMPLE 10.17

Using the values given in Example 10.2 for a waterplane

½ Breadth	SM	Products
2·0	1	2·0
7·3	4	29·2
9·8	2	19·6
10·4	4	41·6
10·6	2	21·2
10·7	4	42·8
10·6	2	21·2
9·9	4	39·6
7·8	2	15·6
4·2	4	16·8
0·2	1	0·2
	30	249·8

Length of base = external multiplier × common interval × sum of internal multipliers

Maximum ordinate = 10·7 × 2 = 21·4 m.

$$\text{Mean ordinate} = \frac{\text{area}}{\text{length of base}} = \frac{\tfrac{1}{3} \times 12\cdot 2 \times \text{sum of functions for area}}{\tfrac{1}{3} \times 12\cdot 2 \times \text{sum of internal multipliers}}$$

SI UNITS — INTERNATIONAL SYSTEM OF UNITS 157

$$= \frac{\text{sum of functions for area}}{\text{sum of internal multipliers}}$$

$$= \frac{249 \cdot 8}{30} = 8 \cdot 37 \text{ m for one side}$$

So that in this case the mean ordinate of waterplane $= 8 \cdot 327 \times 2 = 16 \cdot 65$ m

$$C_w = \frac{A_w}{LB} = \frac{2031 \cdot 7}{122 \times 21 \cdot 4} \text{ from example 10.2}$$

$$= 0 \cdot 778$$

Also $\quad C_w = \dfrac{\text{mean ordinate of waterplane}}{\text{maximum ordinate}} = \dfrac{16 \cdot 65}{21 \cdot 4} = 0 \cdot 778$ as above

As stated under *Volumes and Centroids of Volumes* the area under a curve of immersed areas of sections is a measure of the volume of the ship to the waterline and the longitudinal position of the centroid of the area is the same as the longitudinal centroid of the volume.

The mean ordinate of such a curve is the prismatic coefficient of the ship.

EXAMPLE 10.18.

The ordinates of the curve of areas of immersed sections for a ship, the midship area being treated as unity, are as follows:

Station   A.P. ½ 1   2   3   4   5   6   7   8   9   9½   F.P.
Ordinate 0·03 0·19 0·36 0·69 0·89 0·99 1·00 0·99 0·94 0·74 0·33 0·13   —

The ship has a length of 128 m, breadth 17·7 m, draught 7·9 m and the midship area coefficient is 0·98. Determine the prismatic coefficient and the displacement.

NOTE: AP = after perpendicular; FP = forward perpendicular

Station	Ordinate	SM	Products
A.P.	0·03	½	0·02
½	0·19	2	0·38
1	0·36	1½	0·54
2	0·69	4	2·76
3	0·89	2	1·78
4	0·99	4	3·96
⊗ 5	1·00	2	2·00
6	0·99	4	3·96
7	0·94	2	1·88
8	0·74	4	2·96
9	0·33	1½	0·50
9½	0·13	2	0·26
F.P.	—	½	—
		30	21·00

Common interval $= \dfrac{128}{10} = 12\cdot 8$ m

Mean ordinate $= \dfrac{\text{sum of functions for area}}{\text{sum of internal multipliers}}$

$= \dfrac{21\cdot 00}{30}$

$= 0\cdot 70 = C_p$

$C_b = C_p \times C_m$
$= 0\cdot 70 \times 0\cdot 98$
$= 0\cdot 686$

Displacement $= L \times B \times H \times C_b \times 1\cdot 025$
$= 128 \times 17\cdot 7 \times 7\cdot 9 \times 0\cdot 686 \times 1\cdot 025$
$= 12580$ tonnes [SW]

Also Area of immersed midship section $= B \times H \times C_m$
$= 17\cdot 7 \times 7\cdot 9 \times 0\cdot 98$ m²

So Displacement $= \tfrac{1}{3} \times 12\cdot 8 \times 21 \times 17\cdot 7 \times 7\cdot 9 \times 0\cdot 98 \times 1\cdot 025$
$= 12580$ tonnes as above.

EXAMPLE 10.19.

A ship of 12,000 tonnes displacement has a length of 130 m, breadth 17·5 m, and draught 7·3 m. The prismatic coefficient is 0·85.

Determine the area of the immersed midship section.

$C_b = \dfrac{12000 \times 0\cdot 975}{130 \times 17\cdot 5 \times 7\cdot 3} = 0\cdot 704$

From $C_b = C_p \times C_m$

$C_m = \dfrac{0\cdot 704}{0\cdot 85} = 0\cdot 829$

$\therefore A_m = 17\cdot 5 \times 7\cdot 3 \times 0\cdot 829 = 105\cdot 8$ m²

EXAMPLE 10.20

A ship has a length of 100 m, breadth 14·5 m and a uniform load draught of 5·5 m. Estimate the dimensions of a ship geometrically similar to the above but with only two-thirds the displacement.

In similar ships the coefficients are identical and the displacement varies as L^3. Linear dimensions vary as L.

$\therefore \Delta \propto L^3$ and $L \propto \Delta^{1/3}$
$\therefore L = 100[2/3]^{1/3} = 100 \times 0\cdot 8736 = 87\cdot 36$ m
$B = 14\cdot 5 \times 0\cdot 8736 = 12\cdot 7$ m
$H = 5\cdot 5 \times 0\cdot 8736 = 4\cdot 8$ m

AREAS, VOLUMES, MOMENTS, DISPLACEMENT 159

EXAMPLE 10.21
A ship of 1,800 tonnes displacement in sea water has the undernoted characteristics.
Block coefficient = 0·54; Immersed area of midship section = 29·7 m²: Midship area coefficient = 0·84; Ratio of breadth to draught = 3·5.
Determine the length, breadth and draught of the ship.

$L \times B \times H \times 0·54 = 1800 \times 0·975$

$$L = \frac{1800 \times 0·975}{35·4 \times 0·54} = 92 \text{ m}$$

$$C_m = 0·84 = \frac{29·7}{BH}$$

$\therefore \quad BH = \dfrac{29·7}{0·84} = 35·4 \quad (1)$

$B = 3·5H = 3·5 \times 3·16 = 11·05$ m
Since $B/H = 3·5$
$\quad B = 3·5H$
From (1) $3·5H^2 = 35·4$

$$H = \sqrt{\frac{35·4}{3·5}} = 3·16 \text{ m}$$

So that $L = 92$ m
$\quad B = 11·05$ m
$\quad H = 3·16$ m

Change of Draught due to Change of Density

Legislation requires that all merchant ships shall have maximum draughts assigned to them. These draughts are assessed on the basis that the ships are floating in sea water. However, ships are in many instances loaded while in water which is fresh or nearly so and consequently the draughts are greater than when the ships are in sea water. For a given weight the volume of displacement is greater in freshwater than in sea water due to the lesser density of the fresh water. In view of this the assigned draught in sea water is supplemented by a "fresh water allowance" which when added to the assigned draught gives the draught to which the ship may be loaded in fresh water. This ensures that the assigned draught in sea water will not be exceeded. The "tonnes per centimetre" at the load draught in sea water is used in computing the fresh water allowance.

When a ship passes from sea water to river water the draught increases due to the difference in the water density.

Assume a ship of mass Δ tonnes at a draught H metres in sea water and at a draught of $[H + x]$ metres in fresh water. The volume displaced in sea water is $\Delta/1·025$ m³. Also volume displaced in fresh water is $\Delta/1.000$ m³.

$$\text{Difference in volumes} = \Delta - \frac{\Delta}{1 \cdot 025} = \frac{0 \cdot 025 \Delta}{1 \cdot 025}$$

also

This difference in volume = area of waterplane × difference in draughts
$$= TPC \times 97 \cdot 5 \times x$$

Hence $x \times 97 \cdot 5 \times TPC = \dfrac{0 \cdot 025 \Delta}{1 \cdot 025}$

$$x = \frac{0 \cdot 025 \Delta}{97 \cdot 5 \times TPC \times 1 \cdot 025} \text{ metres}$$

$$= \frac{2 \cdot 5 \Delta}{TPC \times 100} \text{ cm}$$

$$= \frac{25 \Delta}{1000 \, TPC} \text{ cm}$$

At the ports for loading sea-going ships the water is neither completely fresh nor salt and the density may range from 1,000 to 1,025. It is thus desirable to express the increase of draught in general terms for any density with sea water as the basis in the form:

$$x = \frac{\text{displacement} \times \text{difference in density}}{\text{tonnes per cm} \times \text{lesser density}}$$

or

$$x = \frac{\Delta[P_s - P_r]}{TPC \times P_r} \text{ cm}$$

where P_s = density of sea water; P_r = density of river water; TPC = tonnes per cm in sea water; Δ = displacement in tonnes.

For the purpose of assessing the "fresh water allowance" in accordance with the International Convention on Load Lines P_s is taken as 1,025 and P_r as 1,000.
Hence

$$\text{FW allowance} = \frac{\Delta[1025 - 1000]}{TPC \times 1000} = \frac{\Delta}{40 TPC} \text{ cm}$$

where

Δ = load displacement in sea water in tonnes.
TPC = tonnes per cm in sea water at the load condition.

AREAS, VOLUMES, MOMENTS, DISPLACEMENT 161

EXAMPLE 10.22
Determine the change of draught when a ship of 9,930 tonnes displacement and a *TPC* of 16·1 passes from river water at 1,008 to sea water at 1,024 kg/m³.

$$\text{Decrease of draught in cm} = \frac{9930\,(1024 - 1008)}{1008 \times 16\cdot1} = \frac{9930 \times 16}{1008 \times 16\cdot1} = 9\cdot8\,\text{cm}.$$

Flooding

The effect of damage which destroys the watertightness of a ship's hull is often progressive. It depends upon the extent of damage in relation to the ship's internal arrangement and the type and extent of any cargo in the damaged compartments. The extent of flooding in any compartment depends upon the amount of empty space in the compartment.

If a ship consisted of a single compartment then if the hull is pierced the vessel would, in time, be flooded from end to end and ultimately sink. If, however, the ship is subdivided by transverse watertight bulkheads, then when damage occurs, flooding is restricted to the compartment in which the damage takes place. The ship, of course, loses buoyancy due to the damage and will sink deeper in the water but collecting buoyancy from the above water portion of the hull. The provision of a certain minimum freeboard ensures that there is some reserve buoyancy.

The ship not only sinks in the water but also trims by the bow or stern depending upon whether the damaged compartment is forward or aft. After damage, the water level rises inside the ship until the levels inside and outside are identical. Should this level be above the deck at which the bulkheads terminate, then water is no longer restricted to the compartment in which damage has taken place. The water may have free access to 'tween deck spaces and other hold spaces so that the vessel could eventually sink.

It is, therefore, apparent as a first principle that the subdivision arrangements in a ship should be such as to prevent the "bulkhead deck" being submerged when a compartment is damaged and open to sea.

A simple case will be considered here to indicate the principles involved.

EXAMPLE 10.23
Assume a barge, of constant rectangular cross-section, of length 35 m, breadth 12 m, floating at a uniform draught of 3 m. An amidships empty compartment 11·5 m long is pierced so that water has complete freedom of entry.

Determine the draught after flooding.

1) Since the flooded compartment is amidships the barge will sink to a draught t without changing trim. Figure 10–33.

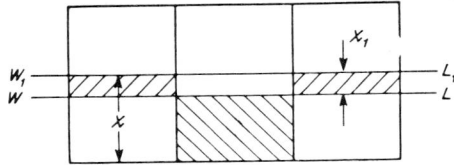

FIG. 10-33—*Effect of flooding on draught—Example 10.23.*

The original immersed volume $= 35 \times 12 \times 3$ m³

Since the mass of the barge is not affected by the damage the above volume must be maintained for the barge to remain afloat.

The intact length after damage $= 35 - 11\cdot5 = 23\cdot5$ m

So that $\qquad 23\cdot5 \times 12 \times t = 35 \times 12 \times 3$
$$t = 4\cdot47 \text{ metres}$$

2) The flooding may be regarded as a loss of buoyancy
The lost buoyancy $= 11\cdot5 \times 12 \times 3$ m³
the lost waterplane area $= 11\cdot5 \times 12$ m²

Intact waterplane area $= (35 \times 12) - (11\cdot5 \times 12) = 23\cdot5 \times 12$ m²
Let sinkage $= t_1$

then $\qquad 23\cdot5 \times 12 \times t_1 = 11\cdot5 \times 12 \times 3$
$$t_1 = 1\cdot47 \text{ m}$$
Final draught $= 3\cdot0 + 1\cdot47 = 4\cdot47$ metres

The general expression is
Gain of buoyancy by intact portion $=$ lost buoyancy

Permeability

The foregoing is on the assumption that the compartment thrown open to the sea is empty so that the volume of water admitted is the volume of the compartment. Actually the volume of admitted water is less due to internal structure such as frames, beams etc. as well as the presence of cargo, stores etc.

The percentage volume of a space that can be flooded is the permeability and is usually designated by the symbol μ.

Thus

$$\text{Permeability} = \mu = \frac{\text{available volume}}{\text{total volume}}$$

A completely empty compartment would have a permeability of 100%.
A completely filled compartment would have a permeability of 0%.

EXAMPLE 10.24

A ship of breadth 15·8 m and TPC 16·1 floats at a uniform draught of 7·3 m. An amidships compartment of length 18 m has a permeability of 0·85. If this compartment becomes open to the sea determine the new draught assuming the ship does not alter trim.

AREAS, VOLUMES, MOMENTS, DISPLACEMENT   163

It is convenient here to use the equivalent empty length of compartment.

Actual length with 0·85 permeability = 18 m.
Equivalent empty length = 18 × 0·85 = 15·3 m.
Total area of waterplane = $TPC \times 97\cdot5$ m^2.
Intact area of waterplane = 97·5 TPC − (15·3 × 15·8) m^2.
Let t = increase in draught in metres.

Then
$t \times$ intact area = volume of buoyancy lost up to original waterline
$t\,[(97\cdot5 \times 16\cdot1) - (15\cdot3 \times 15\cdot8)] = 15\cdot3 \times 15\cdot8 \times 7\cdot3$
$$t\,(1570 - 242) = 1765$$
$$1328\,t = 1765$$
$$t = 1\cdot33 \text{ m}$$
∴ New draught = 7·30 + 1·33 = 8·63 metres

EXAMPLE 10.25

A ship of breadth 16 m and uniform draught 7·3 m has a TPC of 15·5. An amidship compartment of length 18 m, breadth 16 m and of constant rectangular cross-section is bilged and the draught of the ship is then 8·7 m.

Determine the permeability of the compartment.

Area of intact waterplane = $TPC \times 97\cdot5 = 15\cdot5 \times 97\cdot5 = 1511$ m^2.

$$\text{Sinkage} = 8\cdot7 - 7\cdot3 = 1\cdot4 \text{ m}$$

$$\text{Sinkage} = \frac{\text{Volume of lost buoyancy (m}^3\text{)}}{\text{Area of intact waterplane (m}^2\text{)}}$$

$$\therefore 1\cdot4 = \frac{18 \times 16 \times 7\cdot3 \times \mu}{1511 - (18 \times 16 \times \mu)}$$

$$2102\,\mu = 2115 - 403\,\mu$$
$$2505\,\mu = 2115$$
$$\mu = \frac{2115}{2505} = 0\cdot84 = 84\%$$

CHAPTER 11

Transverse Stability

The two principal forces acting upon a freely floating ship are its weight and buoyancy. Other forces, such as wind pressure, act upon it from time to time, and in special circumstances their effect upon stability has to be considered. They do not however, enter into the fundamental problem of ship stability.

The weight and the buoyancy are equal in magnitude; both are vertical forces and for equilibrium counteract each other, so that the following forces act upon the ship when at rest in still water:

1) the weight acts downwards through the centre of gravity (G).

2) the buoyancy acts upwards through the centre of buoyancy (B).

The centre of gravity G is in the same vertical line with the centre of buoyancy B.

Statical Stability

The statical stability of a ship is the tendency the vessel has to return to the upright when inclined from that position.

A ship is said to be in *Stable Equilibrium* if on being slightly inclined from the position of rest she tends to return to that position.

A ship is said to be in *Unstable Equilibrium* if on being slightly inclined from the position of rest she tends to move away farther from that position.

A ship is said to be in *Neutral Equilibrium* if on being slightly inclined from the position of rest she neither tends to return to nor move farther from that position.

If B and G are not in the same vertical line then the two forces of weight and buoyancy, although equal and opposite in direction, are not in the same line and the ship will then turn in the direction of the arrow as shown in Figure 11–1. This is because the moment of the forces acting upon the ship is not zero. If a ship be forcibly inclined and then released, it will roll from side to side for some time and when finally it comes to rest the lines of action of weight and buoyancy will be in the same vertical line.

Transverse Metacentre

Figure 11–2 represents, in section, a ship steadily inclined at a small angle from the upright by some external force. The mass is unchanged

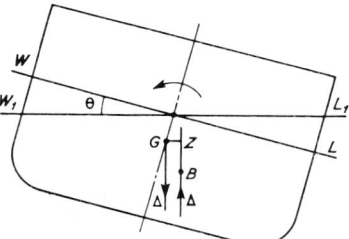

Fig. 11-1—*Heeling forces.*

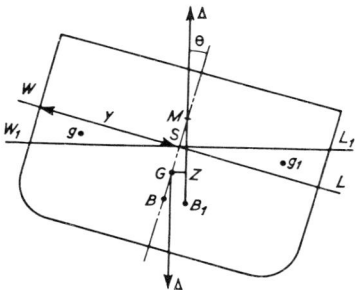

Fig. 11-2—*Heeling forces.*

and consequently the volume of displacement remains the same. It is assumed that no loads on board shift and thus the centre of gravity of the ship is unchanged. However, although the volume of displacement remains the same, the shape of this volume changes and hence the centre of buoyancy will move from its original position.

In Figure 11-2 for the upright position the waterline is WL and when inclined the waterline becomes W_1L_1. The wedge-shaped volume represented by WSW_1, is termed the "emerged" wedge and the volume represented by LSL_1, which has gone into the water is termed the "immersed" wedge. The volume of the emerged wedge and that of the immersed wedge are equal since the ship retains the same volume of displacement. It should be noted that it is only for small angles of inclination that the point S, where WL and W_1L_1 intersect, falls on the middle line of the ship.

When inclined at a small angle from the upright the new volume of displacement has its centroid at say B_1. The upward force of buoyancy must act through B_1 while the weight of the ship acts vertically downwards through G, the centre of gravity of the ship. The intersection of the middle line by the vertical through B_1 at small angles of heel, at the point M is called the *Transverse Metacentre*. GZ is perpendicular to the vertical through B_1. There are thus equal forces acting at a distance from each other of GZ. Such a system of forces is termed a couple and in this case the couple tends to turn the ship back to the upright.

For a ship to be in *Stable* equilibrium for any direction of inclination, the point M must be above G the centre of gravity of the ship. It is found that for all practical purposes in normal merchant ships the point M does not change in position for inclinations up to about 15 degrees. Beyond this it takes up different positions.

For initial stability or stability in the upright the ship is
1) In stable equilibrium if G is below M
2) In unstable equilibrium if G is above M
3) In neutral equilibrium if G and M are coincident

The relative positions of the centre of gravity and the transverse metacentre are extremely important with regard to their effect on the ship's initial stability. The distance GM is termed the transverse metacentric height, and when M is above G then the GM is considered positive. If G is above M the metacentric height is considered negative.

Since for small angles of heel M remains in practically a constant position then for angles up to about 15 degrees.

$$GZ = GM \sin \theta$$

and the couple acting on the ship is

$$\Delta \times GZ \quad \text{or} \quad \Delta \times GM \sin \theta$$

When M is above G this is a righting moment of statical stability. GZ in this case is known as the righting lever of stability. It is the horizontal distance between the lines of action of buoyancy and weight.

$$\Delta \times GZ \quad \text{or} \quad \Delta \times GM \sin \theta$$

is called the moment of statical stability.

This is termed the metacentric method of determining stability but can only be used for small angles up to about 15 degrees.

To Determine the Distance of the Transverse Metacentre above the Centre of Buoyancy (BM)

Assuming that the volume of displacement remains constant and that the transverse inclination of the ship is small, i.e. not greater than 15 degrees then from Figure 11-2

$$BB_1 = BM\theta \text{ where } \theta \text{ is in radians}$$

$$v \times gg_1 = \overline{V} BB_1$$

or $\qquad v \times gg_1 = \overline{V} BM\theta \qquad (1)$

Now
$$v = \int y \frac{y}{2} \theta dL$$
$$= \tfrac{1}{2}\theta \int y^2 dL$$

By substitution in (1)

$$\tfrac{1}{2}\theta \int y^2 dL \times \frac{4y}{3} = \overline{V}\,BM\,\theta$$

$$\therefore \quad BM = \frac{\tfrac{2}{3}\int y^3 dL}{\overline{V}}$$

where v = volume of immersed or emerged wedges
$\overline{V}$ = volume of displacement of ship
gg_1 = distance between centroids of triangular wedges = $4y/3$
y = half-breadth of waterplane
dL = element of length of ship

But
$$\frac{2}{3}\int y^3 dL = I$$

equals transverse moment of inertia of waterplane about the middle-line. (*Note*: See chapter 10 under "Moment of Inertia")

$$\therefore \quad BM = \frac{I}{\overline{V}}$$

The procedure to calculate I has been given in Chapter 10 and illustrated by Example 10.10.

EXAMPLE 11.1.
A ship of 6,000 tonnes displacement has a transverse moment of inertia of the load waterplane of 16,300 m^4.
Determine the transverse BM at the load draught.

$$BM = \frac{I}{\overline{V}} = \frac{16{,}300}{6{,}000 \times 0{\cdot}975} = 2{\cdot}79 \text{ m.}$$

Metacentric Diagram

This is a graph showing the variation in the height of the metacentre (M) with variation in draught. A typical diagram is shown in Figure 11–3. The construction of the diagram can be as follows:

A line is drawn through the origin at 45° and waterlines are drawn to cut this sloping line. At the intersection of the waterline (wl) and the 45° line a vertical is drawn and on this vertical the corresponding height of the metacentre (m) and centre of buoyancy (b) is set up. This is the same as setting off the draughts both horizontally and vertically and allows the position of the metacentre to be seen relative to the waterline. This construction enables the values of KB and BM to be read off separately at any given draught

Now
$$KM = KB + BM$$

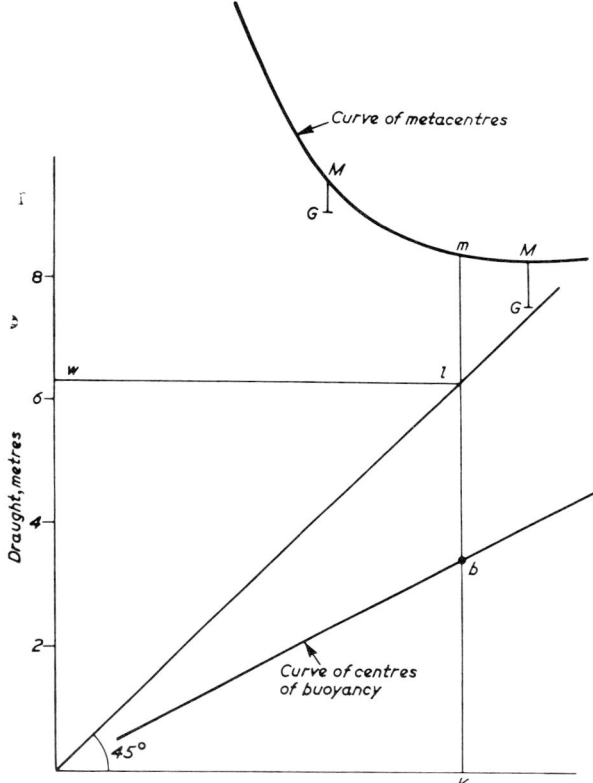

FIG. 11-3—*Metacentric diagram.*

The procedure to calculate the values of *KB* has been shown in Chapter 10 and for *BM* is indicated above.

Approximations to the values of *KB* are given by

1) a first approximation to *KB* at load draught (H)

$$KB = 0 \cdot 52H$$

2) Morrish's Expression

$$KB = H - \frac{1}{3}\left[\frac{H}{2} + \frac{\overline{V}}{A}\right]$$

$$= H - \frac{1}{3}\left[\frac{H}{2} + \frac{LBH\ C_b}{LB\ C_w}\right]$$

$$= H\left[\frac{5}{6} - \frac{C_b}{3C_w}\right]$$

3) A very convenient expression is

$$KB = \frac{C_w}{C_w + C_b} H$$

where $\overline{V}$ = volume of displacement [m^3]
A = area of waterplane [m^2]
C_b = block coefficient
C_w = water plane area coefficient

Metacentric Height.

As stated above the distance between the transverse metacentre M and the centre of gravity G is represented by GM the metacentric height.

If M is above G the metacentric height is considered as positive and if below G is considered as negative.

The metacentric height GM is accepted as the measure of initial stability of a ship at small angles of heel. The GM should for all ships in any condition of loading be positive. In the worst condition of loading the minimum GM should be about 0·3 m.

A ship with a large GM tends to roll quickly and with a jerky motion, whereas a ship with a small GM will roll slowly and easily and with little strain on the hull structure.

Inclining Experiment.

To obtain the metacentric height it is necessary to determine the position of the centre of gravity G. A metacentric calculation consists of two parts. The first is the height of the metacentre above the keel (K) and this is purely geometrical depending only on the form of the ship. The second is the height of the centre of gravity above K and this depends on the hull, machinery mass, loading and free surfaces.

Ship's officers are primarily concerned with the amount of GM in any particular loading condition. The designer is very concerned about the position of the centre of gravity in the light ship or light mass condition, since it is the basis in the calculations for the different conditions of loading.

To determine the position of the vertical centre of gravity an inclining experiment is carried out. This is, in affect, an application of the metacentric method.

The immediate purpose of the experiment is to determine the metacentric height. The ultimate purpose is to obtain the height of the centre of gravity for a definite condition of the ship—the *Light Condition.* A small mass is moved transversely on the deck through a known distance. The resultant angle of heel is measured either by means of two pendulums or by an instrument called a stabilograph which records directly the movement of the ship in degrees. Where pendulums are used they should be as long as possible and of fine wire supporting a heavy bob weight. The deflexion of the pendulum is measured against a horizontal batten. Figure 11–4.

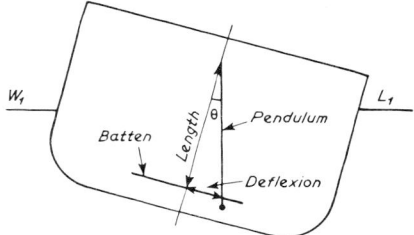

FIG. 11–4—*Inclining experiment.*

Figure 11–5. In the experiment a mass (w) is moved transversely across the deck through a known distance (d). As the mass w is moved across the deck the centre of gravity of the ship moves from G to G_1 such that

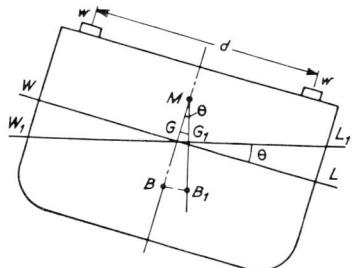

FIG. 11–5—*Inclining experiment.*

$$w \times d = \Delta \times GG_1 \quad \text{where} \quad \Delta = \text{displacement of ship.}$$

At the same time the ship will be inclined through some angle θ and will be in a position of stable equilibrium.

The centre of buoyancy B originally under G will have moved to B_1 so as to be in the same vertical line as G_1. GG_1 is parallel to the movement of w and hence the angle G_1GM is a right angle and

$$GG_1 = GM \tan \theta$$

∴

$$w \times d = \Delta \times GM \tan \theta$$

or

$$GM = \frac{w \times d}{\Delta \tan \theta}$$

also

$$\tan \theta = \frac{GG_1}{GM} = \frac{\text{deflexion of pendulum}}{\text{length of pendulum}}$$

EXAMPLE 11.2

A ship has a displacement of 6,900 tonnes and during an inclining experiment 8 tonnes were moved transversely through 15·5 m. This caused a pendulum of length 9·2 m to deflect 0·22 m. Determine the value of the GM in the inclined condition.

$$\tan \theta = \frac{0 \cdot 22}{9 \cdot 2} = 0 \cdot 0239$$

$$GM = \frac{w \times d}{\Delta \tan \theta} = \frac{8 \times 15 \cdot 5}{6900 \times 0 \cdot 0239} = 0 \cdot 752 \text{ m.}$$

EXAMPLE 11.3.

A ship of light displacement 3,550 tonnes during an inclining experiment recorded on a stabilograph an angle of heel of 1·3 degrees, when 6 tonnes were moved transversely a distance of 10 metres. For the displacement the KM is 8·5 m.

Determine the height of the centre of gravity above the keel.

$$\tan 1 \cdot 3 \text{ degrees} = 0 \cdot 0227$$

$$GM = \frac{w \times d}{\Delta \tan \theta} = \frac{6 \times 10}{3550 \times 0 \cdot 0227} = 0 \cdot 742 \text{ m.}$$

$$KG = KM - GM$$
$$= 8 \cdot 5 - 0 \cdot 74$$
$$= 7 \cdot 76 \text{ m.}$$

EXAMPLE 11.4.

A mass of 80 tonnes is moved 10·5 m transversely across the deck of a ship. The ship's displacement is 8,100 tonnes and the metacentric height 0·54 m. Estimate the resulting angle of heel.

From
$$GM = \frac{w \times d}{\Delta \tan \theta}$$

$$\tan \theta = \frac{w \times d}{\Delta \times GM} = \frac{80 \times 10 \cdot 5}{8100 \times 0 \cdot 54} = 0 \cdot 192$$

from which $\theta = 10°48'$

The principle underlying the inclining experiment is simple but the carrying out of the experiment requires careful attention to ensure a reliable result. Since all subsequent stability calculations are based upon the position of the centre of gravity determined by this experiment it is essential that all the factors which affect its accuracy are considered.

a) A calm day should be chosen.
b) It is desirable that the experiment be made when the ship is nearly, if

not, completed. A list should be made of loads to go on to complete, items to be taken off and items on board that have to be shifted. The vertical and longitudinal positions of their centre of gravity must be recorded.
c) The ship must be floating freely. All mooring wires should be slack and shore gangways removed.
d) All tanks should be empty or pressed full.
e) All workmen, other than those engaged in the experiment, should be sent ashore. Those on board should remain as nearly as possible in a constant position while readings are being taken.
f) All loose items such as derricks, boats, etc. should be secured.
g) The position of the inclining masses should be defined by chalk lines on the deck.
h) At least two plumb-lines should be used one forward and one aft. A whitewood batten should be placed so that its edge is close to the swing of the pendulum. If each successive position of swing is marked and numbered it is possible to check all the figures at the end of the experiment. A heavy plumb-bob at the end of the line is conducive to steadiness and it is usual to have the bob swinging in a container of water in order to damp the oscillations. The pendulum readings could be used as a check on a stabilograph.
i) The draughts must be carefully measured. The draught amidships should also be taken by measuring with a batten the distance from the Plimsoll mark on *each* side to the water level. The hog or sag of the vessel is thus determined, and also the initial list, if any. The density of the water in which the ship is floating should be determined by a hydrometer.

Free Surface

When a tank is completely filled with liquid, no movement is possible and the effect on stability is exactly the same as that of a solid body of the same mass occupying the same space. If, however, the fluid in a tank has a "free" surface such that the slope of the surface can change as the ship is inclined the position of the centre of gravity of the fluid will change as the ship is inclined.

The effect of this free surface is generally expressed as being equivalent to raising the centre of gravity from G to G_1 where G is the centre of gravity of ship and fluid, the fluid being assumed immobile. G_1 denotes the virtual centre of gravity and G_1M is the virtual metacentric height.

Figure 11–6. $GG_1 = GM - G_1M =$ loss of metacentric height.

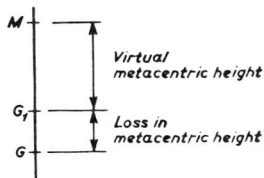

FIG. 11–6—*Effect of free surface liquid on metacentric height.*

If the fluid in the tank is of the same density as that in which the ship floats then

The rise in G or loss of $GM = \dfrac{i}{\overline{V}}$

where i = moment of inertia of the free surface about a fore and aft axis through its centroid.
$\overline{V}$ = volume of displacement of ship.

If the liquid in the tank is different from that in which the vessel is floating then the loss of metacentric height is given by

$$GG_1 \text{ or loss of metacentric height} = \dfrac{P_t}{P_s} \cdot \dfrac{i}{\overline{V}}$$

where P_t = density of liquid with free surface
P_s = density of water in which the ship floats.

Again $\qquad\qquad$ Loss in $GM = \dfrac{i}{\Delta N}$

where Δ = displacement in tonnes sea water
N = number of m^3 occupied by one tonne of the internal fluid.

It is important to note that this free surface effect is quite independent of the depth and volume of the tank and a few centimetres of fluid will have the same effect on initial metacentric height as a large volume. It should also be noted that the effect is independent of the position of the tank in the ship. The tank can be at any height in the ship, at any position along its length and need not be on the middle-line.

EXAMPLE 11.5.

A ship has oil of relative density 0·88 in a tank of length 12 m and breadth 9·5 m. The oil has a free surface. The displacement of the ship is 8,700 tonnes. Determine the loss of metacentric height due to the free surface.

$i = 1/12 \times 12 \times 9 \cdot 5^3 = 857$ m^4 $\qquad\qquad P_t = 0 \cdot 88$
$\overline{V} = 8700 \times 0 \cdot 975$ m^3 $\qquad\qquad\qquad\quad P_s = 1 \cdot 025$

Loss of $GM = \dfrac{0 \cdot 88}{1 \cdot 025} \times \dfrac{857}{8700 \times 0 \cdot 975}$

$= 0 \cdot 086$ m.

EXAMPLE 11.6.

In a ship of displacement 8,100 tonnes with a metacentric height of 0·61 m and KG of 5·65 m, it is proposed to empty a double-bottom ballast (SW) tank at sea. The tank has a length (l) of 13 m, breadth (b) 12 m and depth 1·06 m. Determine the virtual metacentric height when the water level has fallen 0·6 m in the tank.

Water pumped out $= 13 \times 12 \times 0.6 \times 1.025 = 96$ tonnes.
Centroid of this 96 tonnes above base $= 1.06 - 0.3 = 0.76$ m.
Moments about base:

	Mass	Lever	Moment about base
Initial Displacement	8100	5·65	45765
Ballast Ejected	−96	0·76	−73
	8004		45692

New $KG = \dfrac{45692}{8004} = 5.7$ m.

Assuming KM is constant then $KM = KG + GM = 5.65 + 0.61 = 6.26$ m.
Metacentric height $= KM - KG = 6.26 - 5.7 = 0.56$ m.

Loss of GM due to free surface of water in tank $= \dfrac{i}{V}$

$$= \dfrac{1/12 \times 13 \times 12^3}{8004 \times 0.975}$$

$$= 0.24 \text{ m.}$$

∴ Virtual metacentric height $= 0.56 - 0.24$
$= 0.32$ m.

If the tank under consideration of breadth b in the above example has a watertight centre line division then there will in effect be two tanks each of breadth $b/2$ so that from

$$\text{Loss of } GM = \dfrac{i}{V} = \dfrac{2 \times l \times (b/2)^3}{12} \times \dfrac{1}{V}$$

$$= \dfrac{1}{4}\left[\dfrac{lb^3}{12} \times \dfrac{1}{V}\right]$$

$$= \dfrac{1}{4}\dfrac{i}{V}$$

where $i =$ moment of inertia for the undivided tank.
So that the loss of GM is just one-quarter of that for the undivided tank.

EXAMPLE 11.7.

The same tank as in *Example 11.6* but with a watertight centre girder. Thus there are two tanks each of breadth 6 m.

$$\text{Loss of metacentric height} = \dfrac{i}{V} = \dfrac{2 \times 1/12 \times 13 \times 6^3}{8004 \times 0.975}$$

$$= 0.06 \text{ m.}$$

that is, just one-quarter of the previous loss.

EXAMPLE 11.8.

Determine the angle of heel attained when the starboard tank in

Example 11.7 is pumped out to the extent that the water level has fallen 0·6 m. The centre girder is watertight and the port tank remains full.

Moments about base

	Mass	Lever	Moment
Initial displacement	8100	5·65 m	45765
Ballast ejected	−48	0·76 m	−36
	8052 tonnes		45729

New $KG = \dfrac{45729}{8052} = 5\cdot68$ m

Assuming KM unchanged at 6·26 m.
Metacentric height $= KM - KG = 6\cdot26 - 5\cdot68 = 0\cdot58$

Loss of GM due to free surface effect in starboard tank $= \dfrac{i}{V}$

$$= \dfrac{(1/12) \times 13 \times 6^3}{8052 \times 0\cdot975} = 0\cdot0297 \text{ m, say } 0\cdot03$$

∴ Virtual metacentric height $= 0\cdot58 - 0\cdot03 = 0\cdot55$ m.
Taking horizontal moments about the centre-line

	Mass	Lever	Moment
Initial displacement	8100	0	—
Ballast ejected	−48	+3	−144
	8052		−144

Distance of new CG from centre line $= \dfrac{-144}{8052} = -0\cdot0179$ m.

then   Tangent of heel $= \dfrac{0\cdot0179}{0\cdot55} = 0\cdot0325$

∴ $\theta = 1\cdot8°$ to PORT

Effect of Moving Loads Already on Board

If a mass of w tonnes already on board is raised through a distance of h metres then the centre of gravity G of the ship will be raised to G_1 such that $GG_1 = wh/\Delta$ metres where $\Delta =$ displacement of ship in tonnes

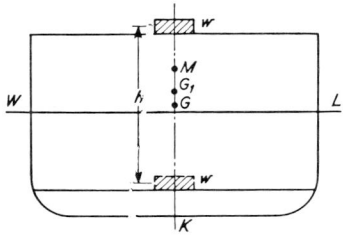

Fig. 11–7—*Effect on stability of vertical movement of load.*

TRANSVERSE STABILITY

Since there is no change in displacement or draught of the ship there will be no movement of the metacentre and consequently the metacentric height will be reduced by GG_1 metres.

If a mass w tonnes already on board is moved transversely a distance d metres then the centre of gravity G of the ship will move transversely in the same direction to G_2 such that

$$\Delta \times GG_2 = w \times d$$

or

$$GG_2 = \frac{w \times d}{\Delta}$$

EXAMPLE 11.9.

A mass of 80 tonnes is lifted from the tank top and placed on deck, the centre of gravity of the mass being raised 8·5 m. If the ship's displacement at that time was 8,000 tonnes determine the reduction in the metacentric height

Loss of metacentric height = rise of $G = GG_1$

$$GG_1 = \frac{w \times h}{\Delta} = \frac{80 \times 8.5}{8000} = 0.085 \text{ m} = 8.5 \text{ cm.}$$

Suspended Weights

Consider a ship lying at a steady angle of heel with a load hanging from a derrick. Whenever the ship moves the load will move; the load, however, hangs vertically below its point of suspension: Consequently the inclination will not be affected if the load be raised till its centre of gravity coincides with its point of suspension. The point of suspension is thus the virtual centre of gravity of the load. Hence the effect of this movable load is precisely that of an equal load having its centre of gravity fixed at the point of suspension.

EXAMPLE 11.10.

A ship of 1,800 tonnes displacement floats upright with a GM of 0·4 m. A derrick lifts a mass of 10 tonnes from the tank top. The point of suspension is 18 m above the centroid of the mass as it lies on the tank top. The mass is then lifted and swung out until the point of suspension is 6 m out athwartships from the mast. Determine the resulting angle of heel.

When the mass is lifted just clear of the tank top its centre of gravity is virtually raised through 18 m. Thus G of ship is raised

$$\frac{10 \times 18}{1800} = 0.1 \text{ m}$$

so that the virtual $GM = 0.4 - 0.1 = 0.3$ m.

When the mass is moved transversely through 6 m the G of the ship is moved transversely

$$\frac{10 \times 6}{1800} = 0.033 \text{ m}$$

Let θ = angle of heel

$$\tan \theta = \frac{0 \cdot 033}{0 \cdot 3} = 0 \cdot 111 \quad \text{and} \quad \theta = 6 \cdot 3°$$

This is the steady angle of heel produced.

NOTE:
When a mass is taken on board or put ashore by derrick it may be necessary to make a correction for the changed position of the metacentre at the new displacement.

The Addition or Removal of Masses

This is a rather more complicated problem since the addition, or removal, of loads from a ship changes the position of the ship's centre of gravity and also changes the draught with a resulting change in the position of the metacentre.

It is possible that at certain draughts the adding or removing of loads may not affect the value of the GM; as instance, a mass taken off the upper deck would lower the ship's centre of gravity and thus increase the GM but this gain could be cancelled by the lowering of M due to the decreased draught.

In dealing with loads of considerable magnitude it is desirable to have complete hydrostatic data for the ship.

To determine the new position of the centre of gravity resulting from a series of alterations in loads being added and removed it is convenient to take moments about the base and arrange the calculation in tabular form as shown in the undernoted example.

EXAMPLE 11.11

A ship of 4,100 tonnes displacement has the centre of gravity 7·3 m above the keel. Loads are added and removed as shown below. Determine the metacentric height in the final condition and the moment of statical stability at 8° inclination. The metacentric diagram gives the KM for the final condition as 7·2 m.

	Loads added		KG	
	800 tonnes		3·7 m	
	600		4·3 m	
	400		6·7 m	
	Load Removed			
	300 tonnes		0·6 m	
		Mass	*KG*	*Moment*
Initial Displacement		4100	7·3	29930
		800	3·7	2960
		600	4·3	2580
		400	6·7	2680
		5900		38150
Removed		300	0·6	180
		5600		37970

Final $KG = \dfrac{37970}{5600} = 6\cdot79$ m.

New metacentric height $GM = KM - KG = 7\cdot2 - 6\cdot79 = 0\cdot41$ m.
Moment of statical stability at $8° = \Delta \times GM \sin \theta$
$= 5600 \times 0\cdot41 \times \sin 8°$
$= 320$ tonnes-metres

Wall-sided Ship

If the sides of a ship are nearly vertical the righting lever GZ can be determined accurately at all angles for which this feature applies. For ships of normal form this can be taken as up to about 15 degrees. The expression for GZ is as follows:

$$GZ = \sin \theta \, (GM + \tfrac{1}{2} BM \tan^2 \theta)$$

where $GZ =$ righting lever; $\theta =$ angle of inclination; $GM =$ original metacentric height; $BM =$ original value.

It should be noted that the second term in brackets—$\tfrac{1}{2} BM \tan^2 \theta$—has been added compared with the usual expression for GZ at small angles, namely $GZ = GM \sin \theta$.

This wall-sided expression $GZ = \sin \theta \, (GM + \tfrac{1}{2} BM \tan^2 \theta)$ is obtained in the following manner.

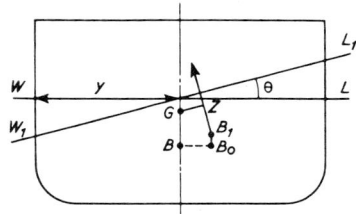

FIG. 11-8—(a) *Righting lever—wall sided ship.*

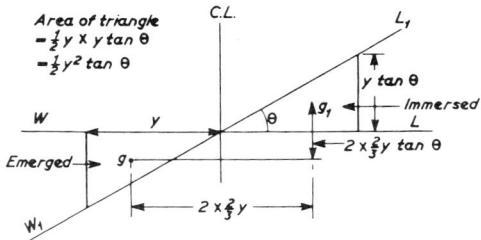

FIG. 11-8—(b) *Righting lever—wall sided ship.*

Let half-breadth at section shown be y and the angle of heel θ, then transfer of moment of triangle horizontally from emerged to immersed side

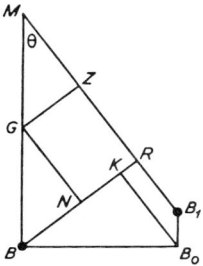

(c) *Righting lever—wall sided ship.*

$$= \tfrac{1}{2} y^2 \tan \theta \times 2 \times 2/3 y$$

and if dl is a small length of the ship in way of section

$$\bar{V} BB_o = \int \tfrac{1}{2} y^2 \tan \theta \times 4/3 y \, dl$$

$$= \tan \theta \int 2/3 y^3 \, dl$$

$$= I \tan \theta$$

where I = moment of inertia of waterplane about S.

$$\therefore \quad BB_o = \frac{I}{V} \tan \theta$$

$$= BM \tan \theta$$

Now transfer of moment of triangle vertically from emerged to immersed side.

$$= \tfrac{1}{2} y^2 \tan \theta \times 2 \times \tfrac{1}{3} y \tan \theta$$

$$\bar{V} B_o B_1 = \int \tfrac{1}{2} y^2 \tan \theta \times 2/3 y \tan \theta \, dl$$

$$= \tfrac{1}{2} \tan^2 \theta \int 2/3 y^3 \, dl$$

$$= \tfrac{1}{2} I \tan^2 \theta$$

$$\therefore \quad B_o B_1 = \tfrac{1}{2} \frac{I}{V} \tan^2 \theta = \tfrac{1}{2} BM \tan^2 \theta$$

To enable the geometry to be more readily followed Figure 11-8(c) is drawn to a larger scale and distorted to give clearness.

BB_0 is drawn parallel to WL and thus at right angles to BM.

B_0B_1 is at right angles to WL and parallel to BM.

B_0K is drawn perpendicular to W_1L_1 and thus parallel to B_1M.

$$GZ = BR - BN$$
$$= BK + KR - BN$$
$$= BB_o \cos \theta + B_oB_1 \sin \theta - BG \sin \theta$$

TRANSVERSE STABILITY 181

Now
$$GZ = BB_o \cos \theta + B_0B_1 \sin \theta - BG \sin \theta$$
$$= BM \tan \theta \cos \theta + \tfrac{1}{2}BM \tan^2\theta \sin \theta - BG \sin \theta$$
$$= [BM - BG] \sin \theta + \tfrac{1}{2}BM \tan^2 \theta \sin \theta$$
$$\therefore \quad GZ = \sin \theta [GM + \tfrac{1}{2} BM \tan^2 \theta]$$

If the GM is *zero* the ship will still have a righting lever.

$$GZ = \tfrac{1}{2} BM \tan^2 \theta \sin \theta$$

If the GM is *negative*, i.e. G above M then the vessel will "loll" over until B_1 and G are in the same vertical line in which case $GZ = 0$ and

$$\tan \theta = \pm \sqrt{\frac{2GM}{BM}}$$

A ship which is initially unstable will not necessarily capsize but heel indifferently to either port or starboard.

At the angle of "loll" the ship has a positive metacentric height $[GM_\theta]$ given by

$$GM_\theta = 2GM \sqrt{1 + \frac{2GM}{BM}}$$

this can also be written as

$$GM_\theta = \frac{2GM}{\cos \theta}$$

where GM = original negative metacentric height; BM = original value.

EXAMPLE 11.12.

A ship has a negative metacentric height of 7·5 cm and a BM of 5 metres. Determine the angle of loll and the positive metacentric height at that angle.

$$\tan \theta = \pm \sqrt{\frac{2 \times 7 \cdot 5}{5 \times 100}} = \pm \sqrt{0 \cdot 03} = \pm 0 \cdot 173$$
$$\therefore \theta = 9 \cdot 8°$$

also

$$GM_\theta = 2GM \sqrt{1 + \frac{2GM}{BM}}$$
$$= 2 \times 7 \cdot 5 \sqrt{1 + \frac{2 \times 7 \cdot 5}{5 \times 100}}$$

$$= 15 \sqrt{1 + 0\cdot03}$$
$$= 15 \times 1\cdot015$$
$$= 15\cdot2 \text{ cm} = 0\cdot152 \text{ m}$$

or
$$GM_\theta = \frac{2GM}{\cos \theta} = \frac{2 \times 7\cdot5}{0\cdot9854} = 15\cdot2 \text{ cm}$$

To determine the angle of heel produced when a mass w is moved transversely a distance d across the deck.

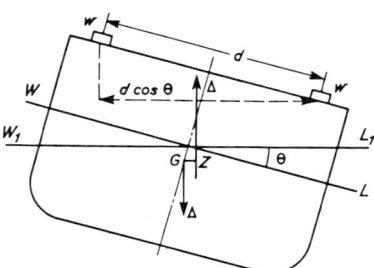

FIG. 11-9—*Example 11.12.*

From Figure 11·9 The horizontal transference of the mass $= wd \cos \theta$ and for equilibrium $wd \cos \theta = \Delta\, GZ$.

Accepting the wall-sided expression to be applicable then
$$wd \cos \theta = \Delta \sin \theta\, (GM + \tfrac{1}{2} BM \tan^2 \theta)$$
In the special case where GM is zero this expression reduces to
$$\tan \theta = \sqrt[3]{\frac{2wd}{\Delta BM}}$$

EXAMPLE 11.13.

In a ship of 8,000 tonnes displacement and zero metacentric height the BM is 3·3 m. Estimate the angle of heel if 20 tonnes are moved 9 m across the deck.

$$\tan \theta = \sqrt[3]{\frac{2wd}{\Delta BM}} = \sqrt[3]{\frac{2 \times 20 \times 9}{8000 \times 3\cdot3}} = 0\cdot2399$$

$\therefore\ \theta = 13\cdot4$ degrees.

EXAMPLE 11.14.

A ship which is wall-sided in way of the waterline has a displacement of 10,000 tonnes, a BM of 3 m and a metacentric height of 0·15 m. Estimate the angle of heel due to shifting transversely a mass of 10 tonnes through 15 metres.

From
$$wd \cos \theta = \Delta \sin \theta [GM + \tfrac{1}{2} BM \tan^2\theta]$$
then
$$wd = \Delta \tan \theta [GM + \tfrac{1}{2} BM \tan^2\theta]$$

For 10 tonnes shifted 15 metres

$$10 \times 15 = 1500 \tan \theta + 15000 \tan^3 \theta$$
$$15000 \tan^3 \theta + 1500 \tan \theta = 150$$
$$100 \tan^3 \theta + 10 \tan \theta = 1$$

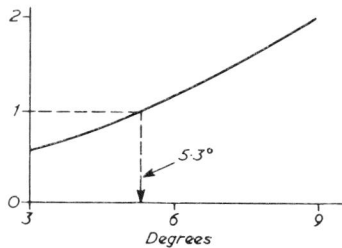

FIG. 11-10—*Example 11.14.*

Let $\theta = 3°$ then *LHS* of expression
$= 0·0143 + 0·524 = 0·538$

Let $\theta = 6°$ then *LHS*
$= 0·1161 + 1·051 = 1·167$

Let $\theta = 9°$ then *LHS*
$= 0·397 + 1·584 = 1·981$

From Figure 11-10. $\theta \doteq 5·3°$

Stability at Large Angles

So far the stability of a ship has been considered only for small angles of inclination. For small angles the statical stability is determined by the metacentric method. When a ship is upright the *GZ* is, of course, zero and thus the *GM* is the only criterion of a tendency to return to the upright at a small inclination. At large angles *GZ* is the direct measure of this tendency.

It is very important to note that

For Small Angles of Inclination
   a) the upright and inclined waterlines intersect on the middle line. Figure 11-2.
   b) the metacentre *M* remains in a constant position.

For Large Angles of Inclination. Figure 11-11.
   a) the upright and inclined waterlines do *not* intersect on the middle line.
   b) the metacentre *M* does not remain in a constant position.

At large angles the determination of the magnitude of GZ involves an appreciable amount of calculation.

Figure 11-11. represents the cross section of a ship inclined at a large angle θ. In the upright the waterline is WL with the centre of buoyancy at B. At the angle θ the waterline is W_1L_1 and the centre of buoyancy is at B_1. The volumes of the immersed and emerged wedges are each equal to v and the centroids of these volumes are g_1 and g respectively. The weight of the ship acts downwards through G and the buoyancy upwards through B_1. GZ and BR are perpendicular to the vertical through B_1.

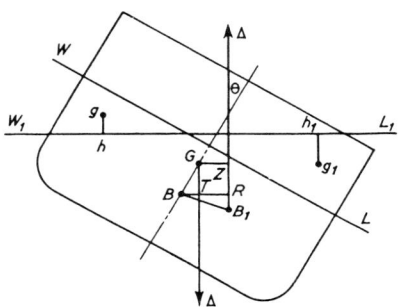

FIG. 11-11—*Stability at large angles of inclination.*

$$\text{Moment of statical stability} = \Delta \times GZ$$
$$= \Delta [BR - BT]$$

By the principle of moments

$$v \times hh_1 = \overline{V} \times BR$$

$$\therefore \quad BR = \frac{v \times hh_1}{\overline{V}}$$

also
$$BT = BG \sin \theta$$
then

$$\text{Moment of statical stability} = \Delta \left[\frac{v \times hh_1}{\overline{V}} - BG \sin \theta \right]$$

This is known as "Atwood's" formula and was developed by George Atwood, F.R.S. in communications to the Royal Society.

As stated the assessment of GZ for large angles of heel involves a considerable amount of calculation and various methods are employed including tabular, integrator and computer. In general what is known as cross curves of stability are prepared.

Cross Curves of Stability

These are curves of GZ plotted on a base of displacement for constant angles of heel. That is, the vessel is assumed to be inclined to some angle θ and the values of GZ are calculated at this angle for a number of displacements between the light and load conditions. Since the position of G depends upon the loading of the ship and is not a fixed position an assumed position for G is adopted for the calculation. Thus, the cross curves so derived are true only for the assumed position of G. The correction necessary when G occupies any other position is easily made as indicated under the heading *Curves of Statical Stability*.

Typical cross curves of stability are shown in Figure 11–12.

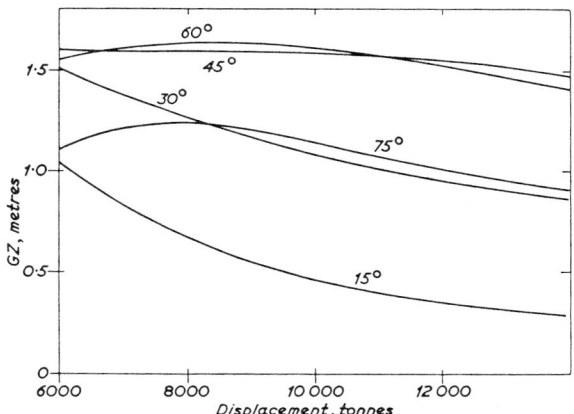

FIG. 11–12—*Typical cross curves of stability.*

Curves of Statical Stability

The curve of statical stability shows for a given displacement and height of centre of gravity above keel (KG) for any angle of heel of the vessel the corresponding value of the lever GZ. Such curves are obtained from a set of cross curves as follows:

Draw a vertical line on the cross curves at the appropriate displacement and measure the ordinate to each of the curves. These can be set up on a base of angles of inclination and thus a curve of statical stability for a given displacement and the assumed KG for the cross curves is derived.

The curve so obtained from the cross curves by direct measurement can be amended to take account of the KG being at some position other than that assumed for the calculation of the cross curves.

In Figure 11–13 the position of the centre of gravity assumed for the cross-curves calculation is G. G_1 and G_2 are the actual positions for two different loadings at the same displacement and inclination.

It is clear that $G_1 Z$ is greater than GZ, and that $G_2 Z$ is less than GZ.

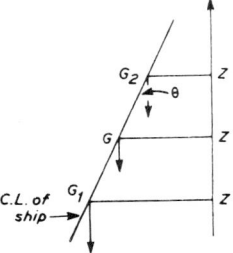

Fig. 11-13—*Assumed position of G for cross curves.*

Thus, if the actual G_1 is below the assumed G then
$$G_1Z = GZ + GG_1 \sin \theta$$
If the actual G_2 is above the assumed G then
$$G_2Z = GZ - GG_2 \sin \theta$$

The following example illustrates the procedure.

EXAMPLE 11.15.

The values of GZ from a set of cross curves for a particular displacement and position of KG are as follows:

Angle	0	15	30	45	60	75	90	degrees
GZ	–	0·18	0·44	0·52	0·41	0·20	−0·06	metres

Plot the curve of statical stability and show the effect of raising the centre of gravity 0·16 m.

Angle	0	15	30	45	60	75	90	at assumed
GZ	–	0·18	0·44	0·52	0·41	0·20	−0·06	KG
$GG_1 \sin \theta$	–	0·04	0·08	0·11	0·14	0·15	0·16	at desired
G_1Z	–	0·14	0·36	0·41	0·27	0·05	−0·22	KG_1

The curve of statical stability is shown in Figure 11-14 from which it will be seen that the reduction in range is about 8 degrees and the reduction in maximum GZ is about 0·1 m.

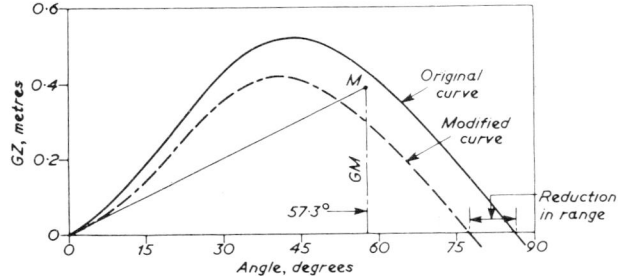

Fig. 11-14—*Curves of statical stability.*

To determine the tangent to the curve of statical stability at the origin, erect at the angle of 57·3 degrees an ordinate representing GM to the same scale as the curve of GZ and join M to 0. In the above Example 11.15 the GM is 0·38 m.

Cross Curves make it possible to obtain easily a statical stability curve for any desired condition of the ship.

A Statical Curve gives
a) the range of stability
b) the value of the maximum lever GZ
c) the angle at which maximum GZ occurs.

Correction of Heel Due to Instability

A ship with a list due to instability will have moved to a position of maximum metacentric height. It is not possible for the master of the ship to increase the height of the metacentre and the sole means by which stability—increased metacentric height in the upright—can be secured is to lower the centre of gravity of the ship. Where empty ballast tanks are available these will afford the simplest means of lowering the ship's centre of gravity.

If the ship is unstable with the ballast tanks full and the cargo holds full then the *only* way to increase the metacentric height is by removing top weight. This in some instances could involve jettisoning cargo.

It is quite wrong to try to correct the heel of a badly listed ship by pumping out ballast water from the low side, if the heel is due to negative metacentric height. The first effect of pumping out such a ballast tank is usually to increase the list. This is because the first water to be removed from a ballast tank extending from the ship's side to the middle line comes from the highest point of the tank and this may well be on the righting side of a vertical through the centre of gravity of the ship.

The correct procedure—assuming there is a reasonable amount of freeboard and the ship's side is intact—is to add ballast on the low side of the ship, since the ballast is then added in the lowest possible position in the ship where it will be of the greatest value for lowering the centre of gravity. The first effect will be to increase the angle of heel and to cause a loss of stability due to the free surface of the water, but this effect is soon cancelled. When the double bottom tank on one side is only partly filled the free surface becomes restricted and the addition of water begins to reduce the angle of heel.

Figure 11–15 shows the starboard double bottom tank partly filled. The tank top has already restricted the free surface of the liquid and as the quantity of water in the tank increases the added water acts through g, the centre of gravity of the layer, which in the diagram is already acting on the port side of G and is both correcting the heel and lowering the centre of gravity.

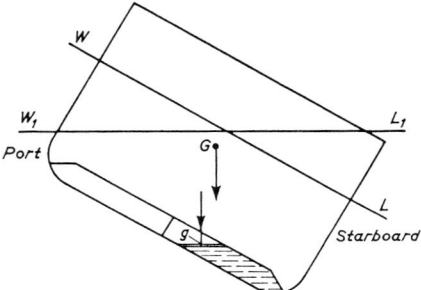

Fig. 11-15—*Starboard double bottom tank, partly filled.*

When some amount of water has been added to the starboard tank, water may then be added to the port tank and the filling of both tanks may proceed simultaneously. By this process, there will be no lurch, and after the initial increase of heel the angle of list will rapidly decrease and the stability will rapidly increase.

CHAPTER 12
Trim

When a ship is inclined about an axis in the fore and aft direction—as in Chapter 11—the considerations involved are termed stability and the change in the position of the ship is measured in degrees. When the ship is inclined about an axis in the athwartship direction—which is considered in this Chapter—the change in the position of the ship is measured by the changes in the draughts forward and aft.

Trim is the difference in draughts forward and aft. If the draught aft is greater than the draught forward the trim is said to be by the stern, if less it is said to be by the head or bow. It is important to know the location of the draught marks as trim is usually referred to between perpendiculars or between marks.

So that with draughts forward and aft as H_F and H_A respectively

$$\text{Trim} = H_A - H_F \text{ assuming } H_A \text{ greater than } H_F$$

and the angle of trim $\theta = \dfrac{H_A - H_F}{L}$

where L is the horizontal distance between the points at which H_A and H_F are measured.

Trim is an important factor in the seaworthiness of a ship and a basic consideration in design. It is the intention of the designer to so arrange the position of the cargo holds and fuel spaces that prior to a voyage the master will have no difficulty in securing a suitable trim when loading without recourse to water ballast.

When a vessel is trimmed through a small angle θ without change in displacement as in Figure 12–1, the trimmed waterline W_1L_1 intersects the original waterline WL at the centre of flotation (F) and equal wedges of immersion and emersion are formed. In effect a wedge of buoyancy aft with centroid at g has been transferred to forward wedge with centroid at g_1.

So that with the transference of wedge of buoyancy (v) there is a transference of the centre of buoyancy of the ship's volume.
Thus

$$\overline{V} \times BB_1 = v \times gg_1$$

A vertical through the new centre of buoyancy B_1 to the new

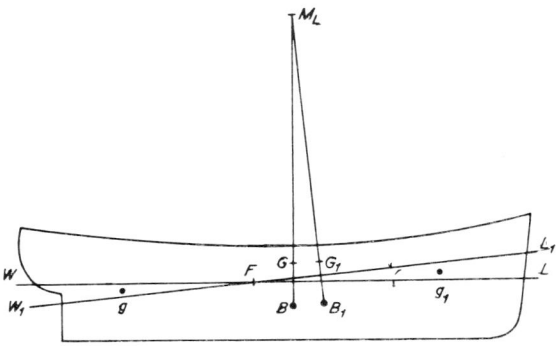

FIG. 12–1—*Ship trimmed through small angle θ.*

waterline W_1L_1 will, when produced, meet the original vertical through B at the point M_L the longitudinal metacentre.

Now $BM_L = I_F/\bar{V}$ where I_F is the longitudinal moment of inertia about the transverse axis through the centre of flotation F.

The calculation of the longitudinal moment of inertia which includes the determination of the area of the waterplane, the position of the centre of flotation, the moment of inertia about amidships and finally, the moment of inertia about the centre of flotation is given in Chapter 10. And

$$BM\ \theta = \frac{v \times gg_1}{\bar{V}}$$

Figure 12–2 represents a wedge of volume forward of the intersection of

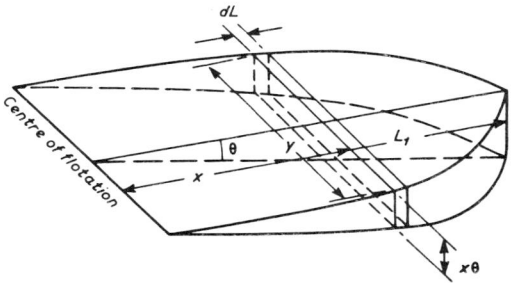

FIG. 12–2—*Wedge of volume.*

WL and W_1L_1. As the angle θ is assumed small it can be taken that the sides of the wedge are vertical. Then

$$\text{volume of element of wedge} = y \cdot x\,\theta\,dL$$
$$\text{volume of whole wedge} = \int_0^{L_1} yx\theta dL = \theta \int_0^{L_1} xy\,dL$$

TRIM 191

Now ydL is the area of an element of the waterplane, and $xydL$ is the moment of the area about the intersection of WL and W_1L_1.

Since the volumes of the wedges are identical the moments of the two must be the same and the intersection of WL and W_1L_1 is the centre of flotation. Again

Moment of element about $F = yx\theta dL \times x = yx^2\theta dL$

Moment of whole wedge about $F = \theta \int_0^{L_1} x^2 y dL$

Now ydL is the area of an element of the waterplane and x^2 is the square of the distance from F and the moment of the whole wedge about F is proportional to the moment of inertia of waterplane about F.

The above also applies to the after wedge, hence

total moment of transference of wedges = moment of forward wedge about F + moment of after wedge about F

$$= \theta \left[\int_0^{L_1} x^2 y dL + \int_0^{L_2} x^2 y dL \right] = \theta I_F$$

where L_2 is distance from F to the aft end. Hence

$$BM \, \theta = \frac{I_F \, \theta}{V}$$

and

$$BM = \frac{I_F}{V}$$

In Example 10.11 the moment of inertia of the waterplane about the centre of flotation is 1,088,700 m⁴. The displacement is 9,250 tonnes. So that

$$BM_L = \frac{1088700}{9250 \times 0.975} = 120.5 \text{ metres}$$

It will thus be observed that the longitudinal BM is greater than the length of the ship. This is the case for most ships.

Change of Trim

The alteration of trim due to the addition, removal or movements of loads or from other causes is called the *Change of Trim*. Suppose a ship is floating at draughts $H_A = 7.92$ m; $H_F = 6.79$ m. Then the trim is 1·13 m by the stern. If loads are added to the ship and the draughts become $H_A = 7.84$ m; $H_F = 7.5$ m the trim will be 0·34 m by the stern.

The change of trim is $1.13 - 0.34 = 0.79$ m. Since the trim by the stern has been reduced the *Change* of trim is by the head.

Change of Trim without Change of Displacement

It has already been stated that when a ship is trimmed through a small angle without change in displacement the trimmed waterline intersects the original waterline at the centre of flotation (F). The total trim is not evenly divided at the draughts forward and aft unless the centre of flotation is at amidships: the change of draught forward and aft will be in proportion to the distance of the centre of flotation from the ends of the ship. In Figure 12–3 F is not at amidships so WW_1 will not equal LL_1. In the figure L_1n is drawn parallel to WL then

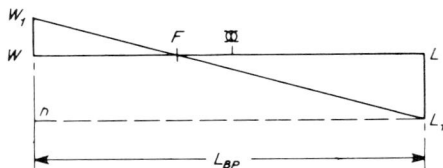

FIG. 12–3—*Change of trim when centre of flotation is not amidships.*

$$W_1n = WW_1 + Wn = WW_1 + LL_1$$

The triangles $W_1 WF$ and W_1nL_1 are similar, from which it follows that

$$\frac{WW_1}{W_1n} = \frac{FW}{nL_1} = \frac{FW}{WL} = \frac{FW}{L_{BP}}$$

∴
$$WW_1 = \frac{FW}{L_{BP}} \times W_1n$$

WW_1 is the alteration of draught at the stern and W_1n is the total change of trim.

Similarly, the alteration of draught forward is given by

$$LL_1 = \frac{FL}{L_{BP}} \times W_1n$$

The change of trim is indicated in the figure by ($WW_1 + LL_1$)

EXAMPLE 12.1

In a vessel of length between draught marks equal to 137 m the centre of flotation is 2·5 m abaft amidships. Assume a change of trim of 1·15 m by the *Stern* without change in displacement. Determine the change in draught forward and aft.

$$FW = \frac{L}{2} - 2·5 = 68·5 - 2·5 = 66·0 \text{ m.}$$

$$FL = 68·5 + 2·5 = 71·0 \text{ m.}$$

Increase of draught aft $= WW_1 = \dfrac{66}{137} \times 1·15 = 0·554$ m $= 55·4$ cm

Decrease of draught forward $= LL_1 = \dfrac{71}{137} \times 1\cdot15 = 0\cdot596$ m $= 59\cdot6$ cm.

Displacement to a Trimmed Waterline.

The displacement calculation is generally made for a series of waterlines parallel to the base line and the displacement curve is drawn using this information. In practice, when it is necessary to obtain the displacement of a ship from a curve or displacement scale the draughts are seldom the same forward and aft and the condition is thus not the same as for the calculation.

The displacement at $W_1 L_1$ is the same as that at WL.

It is usual to accept the centre of flotation F as the waterline corresponding to the arithmetic mean draught $W'L'$ which is $(H_A + H_F)/2$, Figure 12–4. So that to obtain the displacement from the curve the true

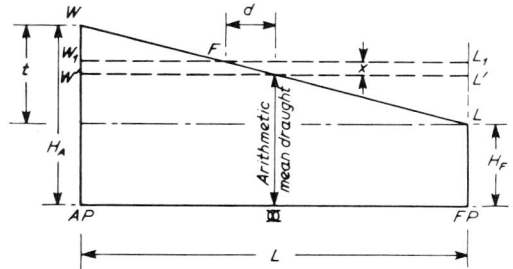

Fig. 12–4—*Displacement to a trimmed water line.*

mean draught is the arithmetic mean draught plus x, where x is the distance between the parallel waterlines $W'L'$ and $W_1 L_1$. In fact x is the depth of the correcting layer.

From the diagram it is evident that

$$\dfrac{x}{d} = \dfrac{\text{total trim }[t]}{\text{length of ship }[L]} \quad \text{where } d = \text{distance of } F \text{ from } \text{⦶}$$

thus $\quad x = \dfrac{td}{L}$

EXAMPLE 12.2.

For a ship of length 122 m at a mean draught on level keel of 7·08 m the centre of flotation is 7·62 m abaft amidships. The measured draughts are H_F 6·56 m and H_A 7·60. Determine the true mean draught.

Measured mean draught $= \dfrac{H_A + H_F}{2} = \dfrac{7\cdot60 + 6\cdot56}{2} = 7\cdot08$ m.

$$t = 7\cdot60 - 6\cdot56 = 1\cdot04 \text{ m.}$$

Correction for trim $= x = \dfrac{td}{L} = \dfrac{1\cdot 04 \times 7\cdot 62}{122} = 0\cdot 065$ m $= 6\cdot 5$ cm

So that true mean draught $= 7\cdot 08 + 0\cdot 065$

$\qquad\qquad\qquad\qquad\quad = 7\cdot 145$ m.

The *TPC* for this ship is, say, 20. The error by reading off the displacement from the curve or scale at the measured mean draught will then amount to $20 \times 6\cdot 5 = 130$ tonnes. This indicates how important it is that the true mean draught be carefully determined.

The determination of the true mean draught for a ship with trim involves adding or deducting the depth of correction layer (x) as shown in Table 12.1.

Table 12.1.

Trim	CF from ⟑	Add or Subtract x to or from Arithmetic Mean Draught
By Stern	Abaft	Add
Stern	Forward	Deduct
By Head or Bow	Abaft	Deduct
Head or Bow	Forward	Add

Moment to Change Trim One cm (MCT 1 cm).

Change of trim is the sum of the changes in draught forward and aft and may be expressed as $L \tan \theta$. Figure 12–5.

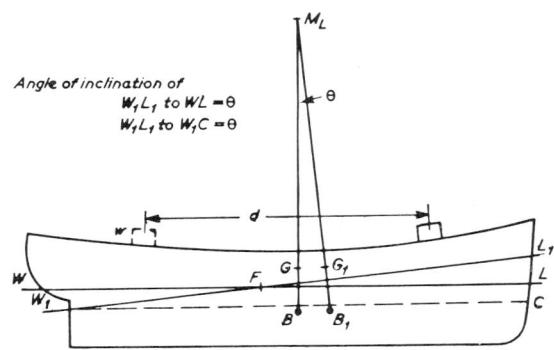

FIG. 12-5—*Moment to change trim.*

Assume ship at waterline *WL* and *G* the position of the centre of gravity of the ship. Suppose a mass of w tonnes is shifted forward a distance d metres. Then $GG_1 = wd/\Delta$; the new waterline is W_1L_1 and the corresponding centre of buoyancy is B_1. The consecutive normals through B and B_1 meet in M_L the longitudinal metacentre.

TRIM 195

The change of trim $= C_{L1} =$ say x

The angle of inclination θ is given by $\tan \theta = x/L$ and also $\tan \theta = GG_1/GM_L$. Thus

$$\frac{x}{L} = \frac{GG_1}{GM_L} = \frac{wd}{\Delta\ GM_L}$$

or $\quad x =$ change of trim in cm $= \dfrac{wdL}{\Delta\ GM_L} \times 100$

and the moment to change trim 1 cm is the value of wd which makes this expression equal to unity

so that $\qquad 1 = \dfrac{wdL}{\Delta\ GM_L} \times 100$

thus wd or $MCT\ 1$ cm $= \dfrac{\Delta\ GM_L}{100_L}$ tonne metre

change of trim in cm $= \dfrac{\text{moment changing trim (tonne metre)}}{MCT\ 1\ \text{cm}}$

The height of M_L varies only with the draught whereas the height of G above the keel varies with the loading and is not constant for each draught. Since the longitudinal BM_L is large when compared with KG it is often assumed in trim calculations that $GM_L = BM_L$.

It is now possible to estimate precisely the changes of draught that will occur when a mass is placed in any position on the ship. The procedure is as follows:

1) Add the mass at the centre of flotation F and determine the increase in draught from the TPC.

2) Move the mass to the desired position and from the change of trim determine the change of draught at each end of the ship.

EXAMPLE 12.3.
A ship of length 107 metres has a displacement of 6 300 tonnes when at draughts $H_A = 6 \cdot 1$ m and $H_F = 5 \cdot 5$ m. At the mean draught of $5 \cdot 8$ m $KM_L = 140$ m and $KG = 5$ m: $LCF = 3$ m forward of amidships. When 100 tonnes of cargo is moved forward a distance of 30 m determine the new draughts.

Trimming moment $= 100 \times 30 = 3000$ tonne metre

$$MCT\ 1\ \text{cm} = \frac{\Delta\ GM_L}{100\ L} = \frac{6300\ (140\text{-}5)}{100 \times 107} = 79 \cdot 5\ \text{tm}$$

196   SHIPS AND NAVAL ARCHITECTURE

$$\text{Change of trim} = \frac{3000}{79 \cdot 5} = 37 \cdot 8 \text{ cm}$$

$$\text{Change of draught aft} = \frac{53 \cdot 5 + 3}{107} \times 37 \cdot 8 = 20 \text{ cm}$$

$$\text{Change of draught forward} = \frac{53 \cdot 5 - 3}{107} \times 37 \cdot 8 = 17 \cdot 8 \text{ cm}$$

so that

	H_A	H_F
original	6·1 m	5·5 m
trim	−0·20	+0·18
New draughts	5·90 m	5·68 m

EXAMPLE 12.4.

For the ship of previous example determine the distance the 100 tonnes of cargo should be moved to bring the vessel to an even keel.

This in effect is to change the trim by $H_A - H_F = 6 \cdot 1 - 5 \cdot 5 = 0 \cdot 6$ m = 60 cm.

$$\text{Change of trim} = \frac{\text{trimming moment}}{MCT \text{ 1 cm}}$$

$$\therefore \quad 60 = \frac{100 \times d}{79 \cdot 5}$$

where d = distance cargo moved in metres

$$\therefore \quad d = \frac{79 \cdot 5 \times 60}{100} = 47 \cdot 7 \text{ metres}$$

EXAMPLE 12.5.

For the same ship as above with a TPC of 12 determine the effect on the draughts of filling a double bottom tank whose capacity is 216 tonnes and centroid 18 m forward of the centre of flotation. Assume $GM_L = 136$ m and $F = 3 \cdot 5$ m abaft amidships.

From the notes given prior to Example 12.3 the procedure is:

1) $TPC = 12$ $\quad \therefore$ Parallel sinkage $= \dfrac{216}{12} = 18$ cm

2) $\quad MCT \text{ 1 cm} = \dfrac{(6300 + 216) \, 136}{100 \times 107} = 82 \cdot 8$ tm

$$\text{Trimming moment} = 216 \times 18 \text{ tm}$$

$$\text{Change of trim} = \frac{216 \times 18}{82 \cdot 8} = 47 \text{ cm by the head}$$

TRIM

$$\text{Trim aft} = \frac{53 \cdot 5 - 3 \cdot 5}{107} \times 47 = 22 \text{ cm}$$

$$\text{forward} = \frac{53 \cdot 5 + 3 \cdot 5}{107} \times 47 = 25 \text{ cm}$$

so that

	H_A	H_F
Original draught	6·1 m	5·5 m
Sinkage	0·18	0·18
Trim	−0·22	+0·25
New draughts	6·06 m	5·93 m

EXAMPLE 12.6.

For the ship as in previous examples determine the draughts forward and aft when 204 tonnes of fuel are removed with centroid 9 m forward of the centre of flotation. Assume *CF* at 3·5 m abaft amidships; *MCT* 1 cm = 79·5 tm and *TPC* = 12.

$$\text{Parallel rise} = \frac{204}{12} = 17 \text{ cm}$$

$$\text{Trimming moment} = 204 \times 9$$

$$\text{Change of trim} = \frac{204 \times 9}{79 \cdot 5} = 23 \cdot 1 \text{ cm}$$

$$\text{Change of draught aft} = \frac{50}{107} \times 23 \cdot 1 = 10 \cdot 8 \text{ cm}$$

$$\text{forward} = \frac{57}{107} \times 23 \cdot 1 = 12 \cdot 3 \text{ cm}$$

so that

	H_A	H_F
Original draught	6·1 m	5·5 m
Rise	−0·17	−0·17
Trim	+0·11	−0·12
New draughts	6·04 m	5·21 m

EXAMPLE 12.7.

A ship is floating at draughts of $H_A = 7 \cdot 35$ m and $H_F = 7 \cdot 00$ m. Cargo is loaded as follows:

60 tonnes	24·5 m forward of amidships
40 tonnes	Amidships
50 tonnes	13·0 m abaft amidships
70 tonnes	21·0 m abaft amidships

The draughts are then $H_A = 7 \cdot 43$ m and $H_F = 7 \cdot 08$ m.

Determine the position of the centre of flotation relative to amidships.

	Load (tonnes)	Centroid from	Moments About ⊗ Aft	Moments About ⊗ Forward
	60	24·5 F	—	1470
	40	⊗	—	—
	50	13·0 A	650	—
	70	21·0 A	1470	—
Total	220		2120 A	1470 F
				2120 A
			Excess =	650 A

thus CF is abaft amidships

	H_A	H_F
Original draught	7·35 m	7·00 m
New draught	7·43	7·08
change	+0·08 m	+0·08 m

i.e. parallel sinkage

∴ $220 \times$ distance of F from amidships $= 650$

∴ F from amidships $= \dfrac{650}{220} = 2 \cdot 95$ m aft

Approximations for MCT 1 cm.

There are occasions when due to incomplete information an approximation to the value of *MCT* 1 cm must be made. The following are in fairly common use.

1) In general terms the longitudinal metacentric height (GML) in the average cargo ship is about 1·2 times the length of the ship. So that

$$MCT \ 1 \ \text{cm} = \frac{\Delta \times GML}{100 L} = \frac{\Delta \times 1 \cdot 2 L}{100 L} \doteq \frac{\Delta}{80} \ \text{tm}$$

where Δ is in tonnes.

2) A convenient expression is

$$MCT \ 1 \ \text{cm} = \frac{7 \ (TPC)^2}{B} \ \text{tm}$$

where $B =$ breadth of ship in metres.

Large Variations of Draught

The foregoing has dealt with the addition or removal of masses which are small when compared with the displacement of the ship. The

TRIM 199

accuracy of the results is based on the assumption that the position of the centre of flotation and the values of *MCT* 1 cm and *TPC* are changed only slightly.

When large masses are involved it is necessary to proceed as follows:
- a) Determine by the metacentre method from the initial draughts the fore and aft position of the centre of gravity (*LCG*).
- b) From the details of the masses and position of the loads added or removed determine the new displacement and *LCG*.
- c) From the hydrostatic curves determine the even keel draught corresponding to the new displacement and the position of its *LCB*.
- d) Estimate by the metacentric method the amount of trim required to move the centre of buoyancy to the same fore and aft position as the new centre of gravity.
- e) The shift of *LCG* or *LCB* for a change of trim can be closely assessed from

$$\text{Shift of } LCB \text{ or } LCG = \frac{\text{change of trim (cm)} \times MCT \text{ 1 cm}}{\text{displacement (tonnes)}}$$

also

$$\text{Change of trim (cm)} = \frac{\text{displacement} \times \text{shift of } LCG}{MCT \text{ 1 cm}}$$

EXAMPLE 12.8.

A ship of length 158 m and displacement 1,300 tonnes floats at draughts of 6·70 aft and 5·30 m forward. At the equivalent even keel draught the *LCB* is 1·6 m aft of amidships and the *MCT* 1 cm is 204 tm. Determine the new draughts after 3,500 tonnes of cargo have been placed on board. The centre of gravity of the cargo can be assumed as 1·85 m. forward of amidships.

Item	Tonnes	LCG (m)	Moments	
			A	F
Ship as floating	13000	3·8 A *	49400	—
Cargo	3500	1·85 F		6480
Total	16500		49400A	6480 F
			6480F	
			42920A	

$LCG = \dfrac{42920}{16500} = 2\cdot 6 \text{ m } A$ * see calculation below

From Hydrostatic curves at 16,500 tonnes displacement
  Draught = 7·46 m   *LCB* = 2·31 m *A*
  LCF   = 5·8 m *A*   *MCT* 1 cm = 235 t m

$$\text{Trim} = \frac{\Delta (LCG - LCB)}{MCT \ 1 \ cm} = \frac{16500 \ (2 \cdot 60 - 2 \cdot 31)}{235} = 20 \cdot 4 \ cm$$

$$\text{Trim aft} = \frac{79 - 5 \cdot 8}{158} \times 20 \cdot 4 = 9 \cdot 45 \ cm, \text{ say } 0 \cdot 095 \ m.$$

$$\text{Trim forward} = \frac{79 + 5 \cdot 8}{158} \times 20 \cdot 4 = 10 \cdot 95 \ cm, \text{ say } 0 \cdot 110 \ m$$

thus

	H_A	H_F
	7·46	7·46
	+0·095	−0·095
New Draughts =	7·56	7·37

or

$$H_A = 6 \cdot 70 \ m$$
$$H_F = 5 \cdot 30 \ m$$
$$\text{Trim} = \overline{1 \cdot 40 \ m} = 140 \ cm$$

Shift of LCG or LCB due to a trim of 140 cm

$$= \frac{140 \times 204}{13000} = 2 \cdot 2 \ m$$

LCB at even keel = $1 \cdot 6 \ mA$

∴ LCG or LCB with $1 \cdot 40$ m trim = $2 \cdot 2 + 1 \cdot 6 = \overset{*}{3} \cdot 8$ m aft of amidships

Adding a Load Without Changing the Draught Aft.

It is sometimes essential to know the position at which a load may be added without any change in the maximum draught when the ship is already trimming by the bow or stern.

The following example indicates the procedure.

EXAMPLE 12.9.

Determine the distance from amidships a mass may be added to the ship without alteration to the draught aft. The ship has a length of 92 m; $TPC = 7$; $MCT \ 1 \ cm = 41$; $LCF = 5 \cdot 3$ m abaft amidships.

Assume mass w is placed at F then sinkage $= \dfrac{w}{TPC}$

Let d = distance mass added forward of amidships
then

$$\text{trim} = \frac{w \ (d + 5 \cdot 3)}{MCT \ 1 \ cm}$$

NOTE: AP to $CF = 46 - 5 \cdot 3$
$$= 40 \cdot 7 \ m$$
$$= a$$

change of draught aft $= \dfrac{w}{TPC} - \dfrac{w(d + 5 \cdot 3)}{MCT} \times \dfrac{40 \cdot 7}{92}$

and this must equal zero

$\therefore \quad \dfrac{w}{TPC} = \dfrac{w(d + 5 \cdot 3)}{MCT} \times \dfrac{40 \cdot 7}{92}$

$\therefore \quad d + 5 \cdot 3 = \dfrac{92 \times MCT}{40 \cdot 7 \times TPC}$

$\therefore \quad d = \dfrac{92 \times 41}{40 \cdot 7 \times 7} - 5 \cdot 3$

$= 7 \cdot 9$ m forward of amidships

The general expression is

Distance load forward of $CF = \dfrac{MCT\ 1\ cm \times L}{a \times TPC}$ where $a = AP$ to CF.

Docking.

When a ship is being dry-docked the draught at the aft end is usually greater than that forward. As the level of the water in the dock is lowered the after end first takes the blocks and until the keel takes the blocks all fore and aft a portion of the weight of the ship is borne by the after block. Immediately prior to the ship taking the blocks fore and aft this portion of the weight is a maximum.

Figure 12-6 shows the ship when about to take the blocks all fore and aft at waterline WL.

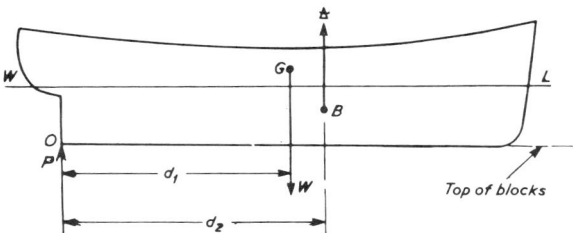

FIG. 12-6—*Ship about to touch down to keel blocks in dry dock.*

Let W = mass of ship floating freely.

Δ = buoyancy at instant when ship about to make contact with blocks all along the keel.

P = pressure on aftermost block at this instant.

then

$\Delta = W - P$

Let d_1 = distance of centre of gravity (G) forward of line of action of P

d_2 = distance of B_1 the centre of buoyancy of Δ forward of P
Then at this instant

$$Wd_1 = \Delta d_2$$
$$= (W - P) d_2$$

$$\therefore \quad P = \frac{W(d_2 - d_1)}{d_2}$$

When the trim by the stern is not excessive the following procedure may be adopted:

$$\Delta\, d_2 = Wd_1 = \text{moment of displacement beneath waterline about 0}$$
$$= \Delta d_2 + (W - \Delta)\, a - Mt$$

where a is distance of the CF of the even keel waterplane forward of the aftermost block and M is the MCT 1 cm; t is the trim in centimetres. Then

$$W - \Delta = \frac{Mt}{a}$$

or

$$P = \frac{Mt}{a}$$

EXAMPLE 12.10.
A ship trimming 0·9 m by the stern is to be docked on level blocks. The length of the ship is 128 m, displacement 4,330 tonnes, LCG is 1·83 m abaft amidships; LCB is 0·92 m forward of amidships; MCT 1 cm = 122 tm. LCF is 60 m from after end.

When the keel is about to make contact with the blocks estimate pressure on block at after end. From

$$P = \frac{W(d_2 - d_1)}{d_2}$$

$$= \frac{4330\,(64\cdot92 - 62\cdot17)}{64\cdot92} = \frac{4330 \times 2\cdot75}{64\cdot92} = 183 \text{ tonnes}$$

or

$$P = \frac{MCT \times t}{a} = \frac{122 \times 90}{60} = 183 \text{ tonnes}$$

Stability on Docking

A ship in dry dock settles down on the keel blocks as the water level falls. Some buoyancy is lost at the surface and to compensate for this a portion of the ship's mass is taken at the keel. The application of a force low down in the ship tends to reduce the stability.

Figure 12-7 illustrates the stability conditions on the assumption that the ship has a small list.

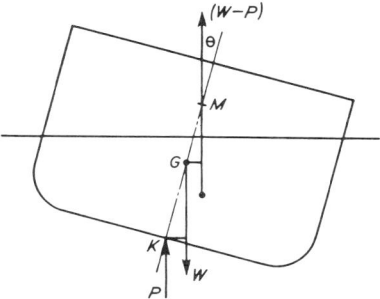

FIG. 12-7—*Stability on docking.*

Taking moments about G

Righting moment $= (W - P) \, GM \sin \theta$

Upsetting moment $= P \cdot KG \sin \theta$

Net righting moment $= (W - P) \, GM \sin \theta - P \, KG \sin \theta$
$= W \, GM \sin \theta - P \cdot GM \sin \theta - P \cdot KG \sin \theta$
$= \sin \theta \, (W \cdot GM - P \cdot GM - P \cdot KG)$
$= \sin \theta \, [W \cdot GM - P \, (GM + KG)]$
$= W \sin \theta \, (GM - P/W \, KM)$

The net righting moment at a small angle θ can be put into the form which will show the virtual loss of metacentric height due to the force (P) at the keel.

Let the expression be $W \cdot GM_\theta \sin \theta$

Then the virtual metacentric height is

$$GM_\theta = GM - P/W \, KM$$

and the loss of metacentric height $= \dfrac{P}{W} \, KM$

M_θ is the point where the resultant upward force W cuts the centre line.

EXAMPLE 12.11.

A ship with trim by the stern and a displacement of 4,330 tonnes is to be docked on level blocks. The KG is 7·64 m and the metacentre at contact is 8·45 m above keel. Determine the virtual metacentric height and the loss of metacentric height.

$GM_\theta = GM - P/W \cdot KM$ $P = 183$ tonnes from Example 12.10.

$$= 0.81 - \frac{183}{4330} \times 8.45$$

$$= 0.81 - 0.36$$

$$= 0.45 \text{ m}.$$

Loss of metacentric height $= 0.36$ m

EXAMPLE 12.12.

A ship of length 150 m with a trim of 130 cm by the stern is to enter a dry dock. If the minimum GM as the vessel takes the blocks all fore and aft is to be 0.40 m determine the minimum allowable GM prior to entering the drydock. The centre of flotation is 2.5 m abaft amidships; MCT 1 cm $= 150$ tonne metre; $KM = 7.5$ m and the displacement $= 6,500$ tonnes. From

$$P = \frac{Mt}{a} \qquad\qquad a = 75 - 2.5 = 72.5 \text{ m}$$

$$= \frac{150 \times 130}{72.5} = 269 \text{ tonnes}$$

Virtual loss of $GM = P/W \cdot KM = \dfrac{269}{6500} \times 7.5 = 0.31$ m

∴ Minimum GM prior to docking $= 0.40 - 0.31$

$$= 0.71 \text{ m}.$$

CHAPTER 13

Watertight Subdivision

When the side of a ship below the waterline is pierced, as by a collision, grounding etc, water pours into the compartment bounded by the bulkheads adjacent to the region of damage and the buoyancy of the ship over the length between the bulkheads is cancelled. A compartment which has been opened to the sea is said to have been bilged.

It is necessary to isolate the flooded volume in order to
a) restrict the loss of transverse stability
b) restrict the damage to cargo
c) restrict trim
d) restrict loss of reserve of buoyancy.

It is essential to have a standard of subdivision such that there is a reasonable chance that the ship will remain afloat under such an emergency.

The requirements for subdivision have been fixed by legislature based on the recommendations of various International Conferences on Safety of Life at Sea.

There are no Government requirements for the number of transverse bulkheads to be fitted in a cargo ship. However, classification society rules specify the number of bulkheads required and this is governed by the length of the ship.

Passenger ships—those which carry more than 12 passengers—must comply with certain standards of subdivision. The method adopted is to determine a line beyond which the ship should not sink and then ascertain the position and length of the compartment which when flooded will cause sinkage to that line. This line beyond which the ship should not sink is known as the *Margin Line*. The relevant terms are now defined.

Bulkhead Deck. This is the uppermost continuous deck to which all transverse watertight bulkheads are carried.

Margin Line. This is a line drawn parallel to and 76 mm, below the upper surface of the bulkhead deck at side.

Permeability. The percentage volume of a space that can be flooded is known as the permeability. In any flooded space the volume of water which can be admitted depends upon whether the space is empty, has cargo, and the nature of the cargo. Some cargoes may fill the space and yet allow more water to enter than will other cargoes, all according to their permeability. For example, a hold filled with sponges and open to the sea could accommodate almost as much water as if the hold were empty; such a cargo would have a very high permeability, say 90 per cent. If a hold were filled with baulks of timber of square cross-section and closely packed, and open to the sea, the only water that could enter would fill the small spaces between the baulks and the small amount absorbed by the timber. Such a cargo would have a very low permeability, say 20 percent. Permeability is usually designated by the Greek symbol μ so that

$$Permeability = \mu = \frac{\text{Available Volume}}{\text{Total Volume}}$$

Floodable Length. The maximum length of a compartment which can be flooded so as to bring a damaged ship to float at a waterline tangential to the margin line.

Curve of Floodable Length. This is a curve which, at every point in its length, has an ordinate representing the length of the ship which may be flooded with the centre of the length at that point, without the margin line being submerged.

There are several methods of determining the position and length of a space which when flooded will cause sinkage to a waterline touching the margin line, including the system prescribed by the International Convention for the Safety of Life at Sea.

The latter is dealt with later but the principles involved can be demonstrated by what is called a direct method in which only the basic hydrostatic data of the ship is required.

Bonjean Curves.

The area of a transverse section of a ship to successive waterlines can be calculated and plotted as a curve showing the variation of sectional area with draught. The curves are frequently drawn on the ships profile at the displacement stations or on a centre line with those for stations in the fore body on the right hand side and for the after body on the left hand side. They enable the displacement and longitudinal centre of buoyancy to be calculated for any waterline, trimmed or even keel. Curves such as these are called Bonjean curves after the Frenchman who first used them. Figure 8–7.

Direct Flooding Calculations.

Figure 13–1 shows the profile of a ship with a waterline WL tangential to the margin line. The purpose of the calculation is to determine the extent and position of the flooding which will bring the ship from waterline W_oL_o to the waterline WL.

Let W_oL_o = waterline of undamaged ship W_o and B_o the corresponding displacement and centre of buoyancy

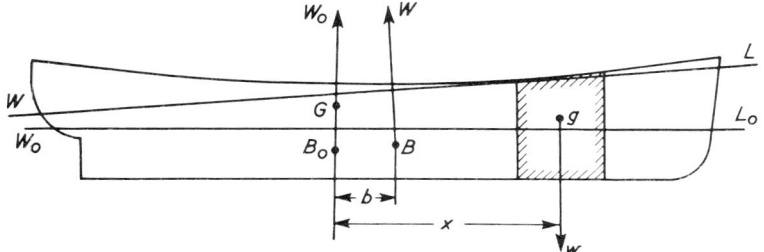

Fig. 13-1—*Waterline tangential to margin line.*

WL = waterline tangential to the margin line with W and B the corresponding displacement and centre of buoyancy.
g = centroid of lost buoyancy w
G = centre of gravity of ship for both conditions

then

Mass of water gaining access to ship *or* the loss of buoyancy $\Big\} = w = W - W_o$

Taking moments about B_o

$$W \times b = w \times x$$

and

$$x = \frac{W \times b}{w}$$

Consequently the extent of the lost buoyancy or of the added weight to the waterline WL and the position of the centroid can be determined. Thus the volume of water admitted is $0.975\,w$ and the total volume of the compartment (v) is given by

$$v = \frac{0.975\,w}{\mu}$$

where μ is the permeability.

The length of the compartment is derived from a curve of areas of immersed sections as follows:

Figure 13-2 shows a portion of such a curve derived from the Bonjean curves at the waterline WL which is tangential to the margin line. The centroid of the added weight or of the lost buoyancy is on the ordinate at A. It is then necessary to determine an area under the curve which will have its centroid on this ordinate at A and also represent the volume $\dfrac{0.975\,w}{\mu}$.

The process is one of trial and error.

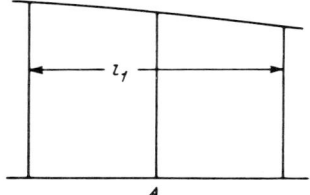

FIG. 13-2—*Portion of curve of immersed area.*

The procedure is to estimate a mean ordinate generally somewhat less than the ordinate at A, say A_1 and the first approximation to the length l_1 of the compartment is given by

$$l_1 = \frac{v}{A_1} \cdot \frac{0.975 \, w \times 100}{\mu \times A_1}$$

where μ is expressed as a percentage.

This length should be laid off so that the middle is on one side or the other of the ordinate at A according to the shape of the curve. The volume and position of the centroid corresponding to the length l_1 can be determined by Simpson's rule using three ordinates; this is used as the basis for a second approximation. Normally the correct length and position are obtained at the second attempt.

The length of the compartment so determined is known as the *Floodable Length* as it is the length in the region considered which may be flooded without making the ship sink beyond the margin line.

By calculations as indicated above for a series of waterlines which are tangential to the margin line at different points throughout the length of the ship it is possible to determine a series of values for the plotting of a set of curves of floodable length as shown in Figure 13–3.

There are usually three curves in the set as there are generally three different average permeability figures along the length of the ship. The floodable length determined by each calculation is set off as an ordinate at the position representing the *Middle of the Floodable Length*. In Figure 13–3, the point A is the mid-length of a compartment of length represented by the ordinate AB. The horizontal and vertical scales used for plotting the curves of floodable length are the same. Thus in the diagram the length of the compartment with mid-length at A is represented by CD where $CA = AD = \frac{1}{2}AB$ and the tangent of the angles BCA and BDA is 2.

It is thus possible to ascertain whether any chosen length of compartment at any position exceeds the floodable length, by plotting the isosceles triangle with the length of the compartment as the base. Thus in Figure 13–3, the length of the compartment represented by EF exceeds the floodable length since the apex G lies above the curve.

The procedure is illustrated by the following example.

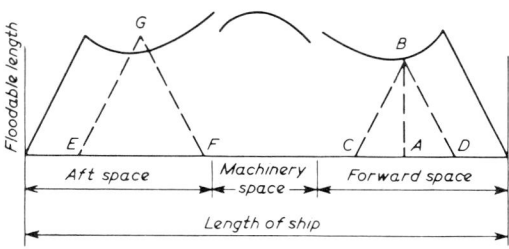

Fig. 13-3—*Curve of floodable length.*

EXAMPLE 13.1.

A ship of length 128 m and displacement 12,200 tonnes with the centre of buoyancy 0·77 m forward of amidships is brought, as the result of damage, to a waterline at which the displacement is 14,200 tonnes and the centre of buoyancy is 7 m forward of amidships. The damage opens to the sea a compartment bounded by transverse bulkheads and with a permeability of 80 per cent. The areas of the immersed sections for the fore body, in the damaged condition, at equidistant stations commencing at amidships, are 139, 148, 158, 162, 139, 84 and 0 m².

Determine the length and position of the damaged compartment.

	Displacement (Tonnes)	*LCB (m)*	*Moment*
Initial condition	12,200	0·77 F	9390
Damaged condition	14,200	7·00 F	99400
Water in compartment	2 000	45mF	90010

$$\text{Volume of compartment} = \frac{2000 \times 0.975}{0.8} = 2435 \text{ m}^3$$

Assume a mean immersed sectional area of 120 m²

then length of compartment $\doteq \dfrac{2435}{120} \doteq 20\cdot 3$ m

Assume bulkheads say 9 m abaft and 11·3 m forward of centroid of flooded compartment.

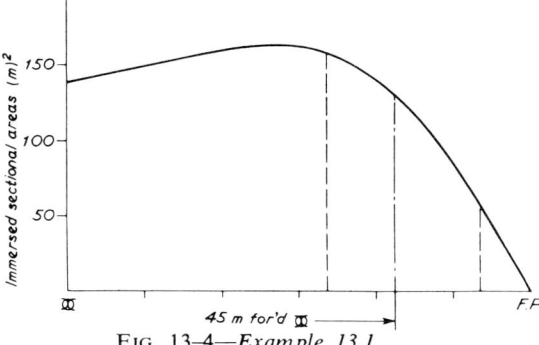

Fig. 13-4—*Example 13.1.*

	Immersed area	SM	f	Lever	f
Aft end	158	1	158	0	—
	124	4	496	1	496
Forward end	65	1	65	2	130
			719		626

Volume of compartment $= 719 \times \frac{1}{3} \times 10 \cdot 15 = 2432$ m³

Centroid of compartment $= \frac{626}{719} \times 10 \cdot 15$

$= 8 \cdot 9$ m forward of aft end of compartment; this more or less satisfies the requirements.

So that Length of compartment $= 20 \cdot 3$ m
 After end of compartment $= 45 - 9 = 36$ m forward of amidships
 Forward end of compartment $= 36 + 20 \cdot 3 = 56 \cdot 3$ m forward of amidships.

EXAMPLE 13.2

A ship of length 158 m has a displacement of 18,500 tonnes and the centre of buoyancy 2·8 m abaft amidships. At a waterline tangential to the margin line the areas of the immersed sections are as follows:

Section AP $\frac{1}{2}$ 1 $1\frac{1}{2}$ 2 3 4 5 6 7 8 $8\frac{1}{2}$ 9 $9\frac{1}{2}$ FP
Area   35 80 115 150 179 213 218 215 210 190 133 94 56 23 m²

Determine the mass of water that has entered the ship and the distance of its centroid from amidships.

	Station	Area (m²)	SM	Function	Lever	Function
	AP	35	$\frac{1}{2}$	18	5	90
	$\frac{1}{2}$	80	2	160	$4\frac{1}{2}$	720
	1	115	1	115	4	460
	$1\frac{1}{2}$	150	2	300	$3\frac{1}{2}$	1050
	2	179	$1\frac{1}{2}$	269	3	807
	3	213	4	852	2	1704
	4	218	2	436	1	436
⊗	5	215	4	860	0	5267 A
	6	210	2	420	1	420
	7	190	4	760	2	1520
	8	133	$1\frac{1}{2}$	200	3	600
	$8\frac{1}{2}$	94	2	188	$3\frac{1}{2}$	658
	9	56	1	56	4	224
	$9\frac{1}{2}$	23	2	46	$4\frac{1}{2}$	207
	FP	—	$\frac{1}{2}$	—	5	—
				4680		3629 F
						5267 A
				Difference =		1638 A

Displacement = 4680 × ⅓ × 15·8 × 1·025 = 25,200 tonnes

$$LCB = \frac{1638}{4680} \times 15\cdot 8 = 5\cdot 53 \text{ m abaft } \text{\textcircled{$\times$}}$$

thus

Final Condition	25,200 × 5·53 A =	139,350
Initial Condition	18,500 × 2·80 A =	51,800
Difference	6,700	87,550

$$\text{Centroid} = \frac{87550}{6700} = 13\cdot 07 \text{ m abaft amidships}$$

thus
Mass of water that has entered the ship = 6,700 tonnes
Centroid of water = 13·07 m abaft amidships.

Draughts at the AP and FP for Waterlines Tangential to Margin Line

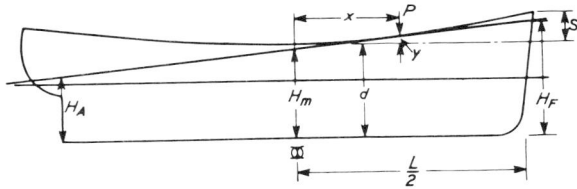

FIG. 13–5—*Draft at after perpendicular and forward perpendicular for water lines tangential to margin lines.*

On the assumption that the lowest point of the margin line is at amidships, that the sheer at the forward perpendicular is S and that the elevation of the margin line is a parabola, then for a waterline tangential to the margin line at P distant x forward of amidships

$$\text{Sheer } y \text{ at } P = \left(\frac{x}{L/2}\right)^2 S$$

and thus

$$\frac{dy}{dx} = 2\left(\frac{x}{L/2}\right)\left(\frac{S}{L/2}\right) = \text{slope}$$

Let d = depth amidships to margin line

$$Hm = \frac{H_A + H_F}{2}$$

Now

$$Hm = d + y - (\text{slope} \times x)$$

$$= d + \left(\frac{x}{L/2}\right)^2 S - 2\left(\frac{x}{L/2}\right)\left(\frac{S}{L/2}\right)x$$

$$= d + \left(\frac{x}{L/2}\right)^2 S - 2\left(\frac{x}{L/2}\right)^2 S = d - \left(\frac{x}{L/2}\right)^2 S$$

Thus
$$H_F = Hm + \text{slope} \times L/2$$
$$= d - \left(\frac{x}{L/2}\right)^2 S + 2\left(\frac{x}{L/2}\right)\left(\frac{S}{L/2}\right) L/2$$
$$= d - \left(\frac{x}{L/2}\right)^2 S + 2\left(\frac{x}{L/2}\right) S$$

Let $f = \dfrac{x}{L/2}$

then $H_F = d - f^2 S + 2fS$
Similarly $H_A = Hm - \text{slope} \times L/2$
$$= d - \left(\frac{x}{L/2}\right)^2 S - 2\left(\frac{x}{L/2}\right) S$$
$$= d - f^2 S - 2fS$$

or
$$H_A = H_F - \text{slope} \times L$$
$$= d - f^2 S + 2fS - 2f \frac{S}{L/2} L$$
$$= d - f^2 S + 2fS - 4fS$$
$$= d - f^2 S - 2fS \text{ as above}$$

EXAMPLE 13.3.

A ship of length 90 m has a depth to the margin line amidships of 9 m. The sheer at the forward perpendicular is 1·8 m. Determine the draughts at the AP and FP for the waterline which is tangential to the margin line at a point (P) 24 m forward of amidships.

Sheer y at $P = (24/45)^2 \times 1·8 = 0·51$ m.

$$\frac{dy}{dx} = 2 \times \frac{24}{45} \times \frac{1·8}{45} = 0·0427$$

$H_F = d - f^2 S + 2fS$ $\qquad\qquad f = 24/45 = 0·533$
$\quad = 9 - (0·533)^2 \times 1·8 + 2 \times 0·533 \times 1·8$
$\quad = 9 - 0·51 + 1·92$
$\quad = 10·41$ m
$H_A = d - f^2 S - 2fS$
$\quad = 9 - 0·51 - 1·92$
$\quad = 6·57$ m

$Hm = d - f^2 S$ $\qquad\qquad$ or $\qquad Hm = \dfrac{10·41 + 6·57}{2}$

$\quad = 9 - 0·51$ $\qquad\qquad\qquad\qquad\qquad = 8·49$ m
$\quad = 8·49$ m

For waterlines tangential to the margin line at points abaft amidships and with sheer at the *AP* being S then:

$$H_A = d - f^2S + 2fS; \quad H_F = d - f^2S - 2fS$$

EXAMPLE 13.4.

A ship of length 158 m has a depth to the margin line amidships of 11 m and sheer at the *AP* of 1·15 m. A waterline is tangential to the margin line at a point 32 m abaft amidships. Assess the draughts at the *AP* and *FP*.

$$H_A = d - f^2S + 2fS \qquad f = \frac{32}{79} = 0\cdot405$$

$$= 11 - 0\cdot164 \times 1\cdot15 + 2 \times 0\cdot405 \times 1\cdot15$$

$$= 11\cdot74 \text{ m}$$

$$H_F = d - f^2S - 2fS$$

$$= 11 - 0\cdot19 - 0\cdot93$$

$$= 9\cdot88 \text{ m}$$

International Conventions for the Safety of Life at Sea (Solas)

As a result of the sinking of the passenger liner *Titanic* in 1912 a Committee was appointed to study the question of the efficient subdivision of ships. They investigated, for a standard form, the effect upon floodable length of variation in block coefficient, freeboard, sheer and permeability. The Committee prepared diagrams based on a standard form to give the floodable lengths as percentages of L for twelve stations and for two permeabilities. The diagram for each station has abscissae of block coefficients and of floodable lengths and ordinates of freeboard ratio and of sheer ratio. The freeboard ratio is f/d where f is the vertical distance amidships from the subdivision load line to the margin line; d is the subdivision draught and is the vertical distance from the moulded base line amidships to the subdivision load line. The sheer ratio, forward or aft, is the ratio of the sheer of the margin line at the *FP* or *AP* measured from the horizontal line through the lowest point of the margin line to the draught.

A specimen diagram is shown in Figure 13–6. To determine the floodable length from such a diagram an ordinate is erected at the appropriate block coefficient, with the top of the ordinate at the appropriate freeboard ratio. From the top of the ordinate a horizontal is drawn to the appropriate point in the curves expressing the effect of the sheer ratio. From the end of the horizontal line an ordinate is dropped to the base line where the desired value of the floodable length, expressed as a percentage of L is read from the scale. The procedure is indicated by lines and arrow heads on Figure 13–6. The floodable lengths so obtained are for 60 per cent and 100 per cent permeability. It is assumed that

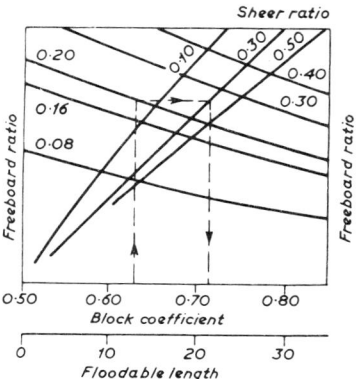

Fig. 13-6—*Determination of floodable length.*

between the permeability values of 60 and 100 per cent the floodable length varies inversely with the permeability.

The Committee considered it desirable to have a system of grading the subdivision of ships and introduced the idea of permissible length being some fraction of the floodable length. The factors whereby floodable lengths are converted into permissible lengths are called *Factors of Subdivision* (F). For any length of ship, above defined limits, there are maximum and minimum possible factors. The maximum value is applicable to ships primarily engaged in the carriage of passengers. The maximum permissible length between bulkheads is obtained by multiplying the floodable length for a point at the centre of the compartment by the factor of subdivision. This factor (F) depends on the length of the ship and the nature of its service as measured by a *Criterion of Service Numeral* (Cs) this numeral is based on the relation between the volume of the spaces allotted to passengers and machinery and the total volume. The greater the volume for passengers and machinery the larger the value of Cs and the lower the value of F.

The work of this committee was the forerunner of the First International Convention for the Safety of Life at Sea. This Convention laid down compulsory regulations for subdivision and it was the duty of the countries signing the convention to ensure that the regulations were observed. The regulations came into force in 1931.

In 1948 an International Conference was held in London to review the operations of the 1931 Convention and make suitable additions and amendments.

Another International Conference was held in London in 1960 to review the decisions of the 1948 Convention. This was deemed necessary in view of the introduction of the nuclear ship and the frequency of the collisions between large ships in certain waters.

The following is a summary of the main decisions of the 1960 International Conference on Safety of Life at Sea.

Application
The 1960 convention applies to all passenger ships—ships which carry more than twelve passengers and to cargo ships of 500 tons gross and over engaged on international voyages. It does not apply to fishing vessels.

Margin Line.
No change was made in this. The 76 mm provision was considered adequate to secure a sufficient reserve of freeboard and stability after damage.

Permeability.
The machinery space permeability was increased from a basic 80 per cent to 85 per cent.

Subdivision Forward
In ships 100 metres in length and upwards, one of the main transverse bulkheads abaft the forepeak shall be fitted at a distance from the forward perpendicular which is not greater than the permissible length.

Two Compartment Standard.
Where the required factor of subdivision is 0·50 or less, the combined length of any two adjacent compartments shall not exceed the floodable length.

Damaged Stability.
Where the administration considers the range of stability in the damaged condition to be doubtful, it may require investigation.

Levelling Arrangements.
In the case of symmetrical flooding there shall be a positive residual metacentric height of at least 0·05 m as calculated by the constant displacement method.

Double Bottoms.
In ships 76 metres in length and upwards a double bottom shall be fitted amidship and shall extend to the fore and after peak bulkheads or as near to there as practicable.

Stability Information.
The inclining experiment for tankers and ore carriers may be dispensed with if the ship's proportions and arrangements clearly

indicate that sufficient metacentric height will be available in all probable loading conditions.

Radio.

Cargo ships of 300 tons gross and upwards but less than 1,600 tons gross must have either radio telegraph or radio telephone.

A typical set of curves of floodable and permissible lengths is shown in Figure 13-7 with the actual subdivisions indicated. All the triangles marking the actual lengths of compartments lie below the curves of permissible length and thus satisfy the regulations.

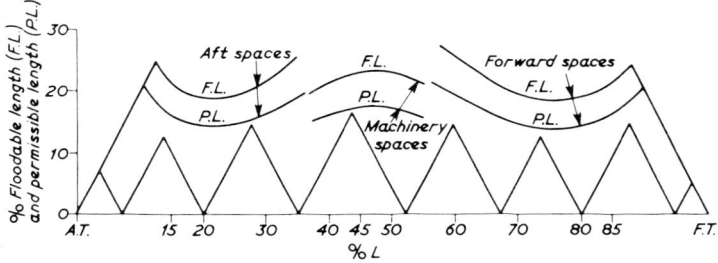

FIG. 13-7—*Curves of floodable length and permissible length.*

CHAPTER 14
Strength of Ships

Longitudinal Bending

As stated in Chapter 5 a ship may be regarded as a loaded beam or girder. In a seaway the ship is subjected to both static and dynamic forces which cause it to bend in a longitudinal vertical plane. For the purposes of structural design and for comparison between ship and ship the problem is considered as a static one, so that it resolves into the ship being poised statically on a wave and the resulting forces and moments acting on the ship calculated.

In order to determine the forces acting on a ship treated as a girder it is necessary to determine the distributions of the weight and the buoyancy. The total weight must equal the total buoyancy and the fore and aft position of the centre of gravity must be in the same athwartship section as the centre of buoyancy.

For the purpose of investigating the longitudinal bending of a ship certain assumptions are made. The calculations are, in general, carried out for two standard conditions, *Hogging* and *Sagging* as shown in Figures 5–4 and 5–5. The assumptions are as follows:

    a) the ship is head on to the waves and is poised statically on a wave;
    b) the wave has a trochoidal profile of length equal to the length of the ship and a height as described below.
    c) the wave crest is at amidships for the hogging condition;
    d) the wave crest is at the ends for the sagging condition.

A long accepted practice has been to take the height of the wave (h) as 1/20th of the length in strength calculations. However, observation of sea waves has shown that longer waves tend to be less steep than shorter waves and this led Lloyd's Register to suggest that a better approximation to the height would be

$$h = 1 \cdot 1 \sqrt{L} \quad \text{for } L \text{ in feet}$$

For metric units this becomes $h = 0 \cdot 607 \sqrt{L}$
which gives $h/L = 0 \cdot 607/\sqrt{L}$ instead of $h/L = 1/20$ as formerly adopted.

Later Murray of Lloyd's Register suggested that wave height should vary as $L^{0 \cdot 3}$ instead of $L^{0 \cdot 5}$ and Muckle has derived from data that the wave height can be written

$$h = 1 \cdot 632 L^{0 \cdot 3} \quad \text{where } L \text{ is in metres}$$

The wave heights derived from these three expressions are given in Table 14.1.

Table 14.1.

L (metres)	L/20	$0.607 \sqrt{L}$	$1.632 L^{0.3}$
60	3	4·71	5·57
90	4·5	5·76	6·29
120	6·0	6·66	6·86
150	7·5	7·44	7·33
180	9·0	8·15	7·75
210	10·5	8·80	8·12
240	12·0	9·41	8·45
270	13·5	9·98	8·75
300	15·0	10·52	9·03

It is apparent that there are considerable differences in the heights obtained from the expressions and consequently considerable differences in the final result of any calculation made therefrom. However, the results obtained from longitudinal strength calculations are considered very largely on a comparative basis so that if the same expression for wave height is adopted when making a comparison between ships the actual height assumed is not very important.

A trochoid is a curve produced by a point at radius r within a circle of radius R rolling on a flat base. The equation to a trochoid with respect to the axes as shown in Figure 14-1 is

$$x = R\theta - r \sin \theta$$
$$Z = r (1 - \cos \theta)$$

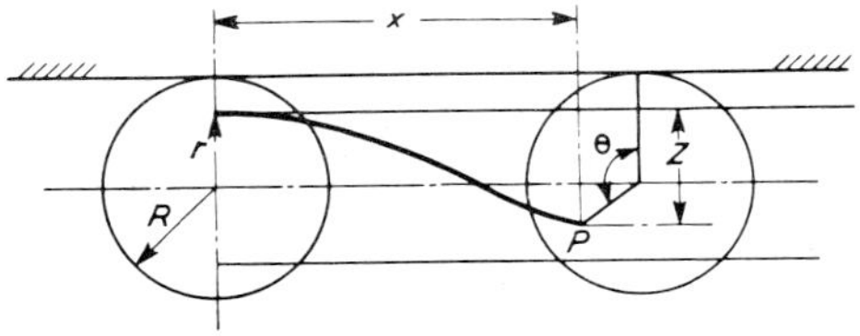

Fig. 14-1—*Trochoid*.

The radius R of the circle is related to the length of the wave since when the circle has made one revolution it must have moved one wave length. Hence

$$L = 2\pi R \quad \text{or} \quad R = \frac{L}{2\pi}$$

The radius r is simply half the height of the wave; $r = h/2$.

STRENGTH OF SHIPS

By the adoption of a trochoidal wave of height $0·607\sqrt{L}$ as a standard wave in the comparative longitudinal strength this wave has the equation

$$x = \frac{L}{2\pi}\theta - \frac{0·607\sqrt{L}}{2}\sin\theta$$

$$Z = \frac{0·607\sqrt{L}}{2}(1 - \cos\theta)$$

To draw the trochoid for the purpose of the strength calculation it is only necessary to set out the length of the wave and divide this into a convenient number of equal parts, say 12. Each interval of length then represents a rotation of the rolling circle of 30 degrees so that if lines of length r are drawn at each of the intervals and inclined at angles to the vertical which increase by 30 degrees for each interval, the necessary points will be secured through which the trochoid may be drawn.

As stated above two conditions are investigated in the longitudinal strength of the ship.

a) *Hogging* with wave crest amidships;
b) *Sagging* with wave crest at the ends.

With a wave crest at amidships the greatest bending moment will occur with a concentration of weight towards the ends. The ship will hog, the deck being in tension and the bottom in compression.

With crests at the perpendiculars the greatest bending moment on the structure will occur when there is a concentration of weight at amidships. The ship will sag with the deck in compression and the bottom in tension.

The two conditions are indicated in Figure 14-2 where the hatched area indicates the positions of the loads.

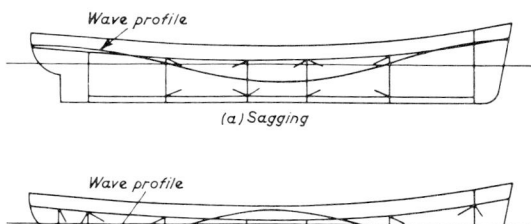

FIG. 14-2—*Sagging and Hogging.*

The strength calculations are generally based on conditions of loading approximating to deep load condition with the following assumptions:
Hogging: Fuel and water carried amidships have been consumed. Any water ballast necessary is carried at the ends of the ship.

Sagging: Fuel and water carried amidships are on board. Any water ballast necessary is carried amidships.

It is not possible to generalize on the type of loading that will produce the greatest bending moment in either hogging or sagging. It is difficult in some ship types to state which load distribution should be adopted and frequently with tankers it may be necessary to investigate a number of conditions in order to determine the worst.

Distribution of Buoyancy

The total upward force on a cross-section of the ship one metre in length is given by the product of the density of the water and the immersed volume of the section. Thus if A is the mean immersed area of the section in m^2 and the density of the water is $1 \cdot 025$ tonne/m^3 the

$$\begin{aligned} \text{upward force} &= A \times 1 \times 1\cdot025 \text{ tonne/m or} \\ &= A \times 1 \times 1025 \times 9\cdot81 \text{ N/m}. \end{aligned}$$

A curve of buoyancy per metre of length can thus be drawn on a base of length of ship. The total buoyancy is represented by the area under the curve and the fore and aft centroid of the area enclosed by the curve represents the longitudinal centre of buoyancy of the ship.

Buoyancy and Balance

The wave profile is drawn on tracing paper and placed over the profile of the ship on which the Bonjean curve has been drawn in at each ordinate. Figure 14-3 shows a trial wave position and one ordinate

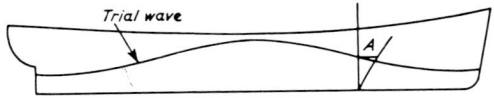

FIG. 14-3—*Trial wave*.

with its Bonjean curve. For equilibrium it is essential to place the wave at a draught and trim such that

　a) the displacement equals the weight and

　b) the centre of buoyancy lies in the same transverse plane as the centre of gravity.

The immersed areas at each ordinate can be read from the Bonjean curves where the wave profile cuts the ordinate. These areas are then subjected to integration—usually a Simpson's rule—to give displacement and the position of the *LCB*. The position of the wave to meet the two conditions (a) and (b) stated above can be determined by trial and error. This can usually be achieved not later than the third attempt.

Having secured a balance, the curve of buoyancy per metre can be drawn as in Figure 14–4.

A very convenient way of obtaining the buoyancy curve is given by W. Muckle in the February issue of *The Shipbuilder*, Vol. 61. By this simple method, the correct displacement and position of the centre of buoyancy can be obtained without having to make several estimates of the position of the wave.

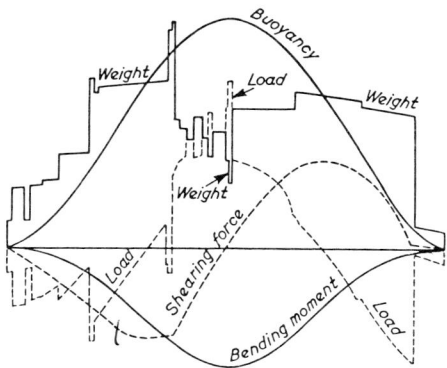

FIG. 14–4—*Curve of buoyancy per metre.*

The Weight Curve

The other item involved in the determination of the forces and moments acting on the ship is the distribution of the load throughout the length of the ship. This is more involved than that of the buoyancy. The determination of a diagram of mass per unit of length is a rather laborious process. The total mass of the ship in any condition of loading can be divided into hull, propelling machinery and deadweight; the latter consisting of cargo, fuel, water, stores, ballast etc.

Because of the labour involved in assessing the distribution of *Hull* mass with reasonable accuracy many strength calculations are carried out on an approximation to the curve of *Hull* mass. There are a number of such approximations and one that is often used in preference to the more detailed approach is that given by Biles, as follows:

If W is the *Hull* mass and the length of the ship is L and the length is divided into three equal parts with ordinates as in Figure 14–5 then area of the diagram =

$$\frac{W}{L} \frac{L}{3} \left\{ \left(\frac{0 \cdot 653 + 1 \cdot 195}{2} \right) + 1 \cdot 195 + \left(\frac{1 \cdot 195 + 0 \cdot 566}{2} \right) \right\}$$

$$= W(0 \cdot 3080 + 0 \cdot 3983 + 0 \cdot 2935)$$
$$= W$$

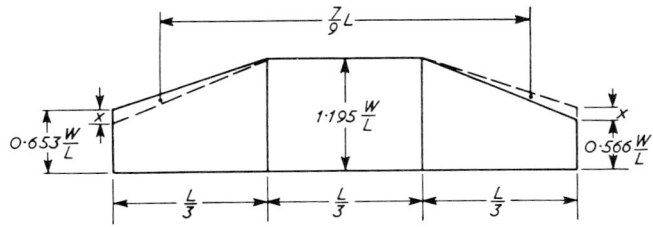

FIG. 14-5—*Distribution of hull mass—approximate method of assessment.*

The centroid of the diagram Figure 14–5 is at 0·0056 L abaft amidships.

The centroid can be moved from this position by transferring a triangle from one trapezium to the other as shown by the dotted lines. The shift of the centroid of the triangle is $7/9L$. Let x be the end ordinate of the triangle to be shifted then,

$$\text{moment of shift} = \text{area of triangle} \times 7/9L$$

$$= x \times \frac{L}{6} \times \frac{7L}{9} = \frac{7}{54} \times x \times L^2$$

Shift of centroid of diagram = moment of shift ÷ W

$$= \frac{7 \ x \ L^2}{54 \ W}$$

$$\therefore \quad x = \frac{54W \times \text{shift of centroid desired}}{7L^2}$$

To the diagram of *Hull* mass it is necessary to superimpose machinery, fuel, water, cargo, etc. in their correct positions and mass per unit length. For items such as fuel, water, cargo, cross-sectional area values are generally available for the compartments to which these items are allocated and the mass per unit length at any point will be proportional to the cross-sectional area at that point.

If a compartment has a volume V cubic metres and contains W tonnes, the length being l metres, then

$$\text{Mass per metre} = \frac{A}{V/l} \times \frac{W}{l} = \frac{AW}{V}$$

Where A is the cross-sectional area at the point considered
or
If w is the mass per cubic metre of the substance in the space considered then mass per metre $= Aw$.

When the diagram is complete the area of the diagram should then equal the total displacement for the condition under consideration and its centroid should correspond to the longitudinal centre of gravity of the ship. Figure 14-4.

Load Curve

The load on the structure at any point is the difference between the mass per unit length (w) and the buoyancy per unit length (b) so that Load on structure $= b - w$. These differences can be plotted as a load curve as shown in Figure 14-4. As the ship is in static balance the areas above and below the base line must be equal.

Shearing Force Curve

If the load curve is integrated then the shearing force (SF) on the structure is obtained, so that

$$SF = \int (b - w) \, dx \qquad \text{Fig. 14-4.}$$

This can be carried out in several ways as indicated below.

Bending Moment Curve

By integrating the shearing force curve the bending moment (BM) on the structure is obtained, since

$$BM = \int SF \, dx \qquad \text{Fig. 14-4.}$$

Methods for integration are given below.

Integration for SF and BM

The load curve may be split up into a number of sections and the area of each of these sections obtained by tracing round with a planimeter or integrator. The shearing force at any point is then simply obtained by adding together the areas of the individual sections up to the point under consideration. The results so obtained when plotted give the shearing force curve.

The shearing force curve can then be divided into a number of sections and the areas obtained in the same way by planimeter or integrator. The sum of these areas then gives the bending moment from which the bending moment curve can be drawn.

The area of each section can also be obtained by the product of the length of each section and its mean ordinate as is used in Example 14.1.

The integraph as mentioned in Chapter 10 is an instrument which, when a pointer is traced round a curve, a pen or pencil draws the integral of that curve. So that if the pointer is traced round the load curve the shearing force curve is automatically drawn out. Again if the pointer is traced round the shearing force curve the bending moment curve is drawn out.

A tabular method of integration is now frequently used in this calculation. Let the length of the ship be divided into a number of equal parts of length l and let b and w be the mean ordinates of the buoyancy and weight curves for any part. The integral of the load curve for this part is then given by $(b - w)l$ and the shearing force at any point will be

$$SF = \Sigma(b - w)l$$
$$= l\,\Sigma(b - w)$$

Thus to obtain the shearing force at any point take the difference between the buoyancy and weight curves for each part, sum these for all parts, and multiply by l. To obtain the bending moment take the mean values of the shearing force for each part and sum these for all parts and multiply by l. Thus, if F_m is the mean value of the shearing force for any part then the bending moment is given by

$$BM = l\Sigma F_m$$

Table 14.2 gives the layout of such a tabular method of integration and is applied to Example 14.1 to indicate the procedure.

Table 14.2.

1	2	3	4	5	6	7
Part	Mean ord. of buoy. curve	Mean ord. of weight curve	Difference	Sum SF	Mean ord. from Col. 5	Sum BM
FP–1	10·6	73	−62·4	−62·4	31·2	31
1–2	34·8	73	−38·2	−101	82	113
2–3	64·7	73	−8·3	−109	105	218
3–4	97·0	73	+24·0	−85	97	315
4–5	125·0	73	+52·0	−33	59	374
5–AP	140·0	73	+67·0	+34	—	374

Shearing Force = sum from Col. 5 × l

Bending Moment = Sum From Col. 7 × l^2

For example the shearing force at Station 2 = 101 × 10·7
 = 1080 tonnes = 10·6 MN
the bending moment at Station 3 = 218 × 10·7²
 = 25,000 tonne metres
 = 24·54 MN metres.

STRENGTH OF SHIPS   225

Summary of Curves

The following relationships should be noted:
1) the area under the weight curve and that under the buoyancy curve are equal;
2) the centroids of the areas of weight and buoyancy are in the same athwartship section;
3) for the load curve the areas above and below the base line are equal;
4) the maximum values of the shearing force occur where the load curve crosses the base line;
5) the maximum bending moment occurs where the shearing force curve crosses the base line;
6) the shearing force and bending moment curves must close. The value at the ends is zero.

It is not always essential to complete the curves of shearing force and bending moment as the concern is generally with maximum values.

At any point the shearing force is given by the difference between the weight and the buoyancy up to that point. Also the bending moment at any point is the difference between the moment of weight and the moment of buoyancy about the point concerned. See Example 14.1 and alternative.

EXAMPLE 14.1.

The ordinates of the curve of buoyancy for the fore body of a ship at stations 10·7 m apart commencing from the forward perpendicular are 0, 21·7, 48·3, 81·0, 113·0, 136·5 and 143·0 tonnes per metre of length of vessel. The weight distribution throughout the fore body is uniformly 73·0 tonnes per metre. Draw the curves of load, shearing force and bending moments and state the maximum values for the SF and BM and the values at amidships.

Figure 14–6 shows the curves.

Area under the load curve gives the SF.

From FP to 1 = 10·7 × 61   = −652   to 1 =   652 tonnes
       1–2 = 10·7 × 40   = −428   2 = 1,080 tonnes
       2–3 = $\dfrac{8·56 \times 12·4}{2·14 \times 4}\Big\}$ = − 97   3 = 1,177 tonnes
       3–4 = 10·7 × 25   = +268   4 =   909 tonnes
       4–5 = 10·7 × 55   = +590   5 =   319 tonnes
       5–$\text{\textit{\foreignlanguage{}{⦻}}}$ = 10·7 × 66·2 = +709   ⦻ =   390 tonnes

Area under the SF curve gives the bending moment.

From FP to 1 = 10·7 ×   350   = − 3,740   to 1 = 3,740 tonne metre
       1–2 = 10·7 ×   910   = − 9,800   2 = 13,540 tonne metre
       2–3 = 10·7 × 1,130   = −12,100   3 = 25,640 tonne metre
       3–4 = 10·7 × 1,085   = −11,660   4 = 37,300 tonne metre
       4–5 = 10·7 ×   640   = − 6,900   5 = 44,200 tonne metre
       5–⦻ = $\begin{matrix}5·3 \times 160\\5·4 \times 195\end{matrix}\Big\}$ = +   200   ⦻ = 44,000 tonne metre

Also the SF at Station 2 = area under weight curve — area under buoyancy curve

$= 2 \times 10\cdot7 \times 73 - [\tfrac{1}{3} \times 10\cdot7 (0 + 4 \times 21\cdot7) + 48\cdot3]$
$= 1{,}562 - 482 = 1{,}080$ tonnes, as above
$= 10\cdot6$ MN

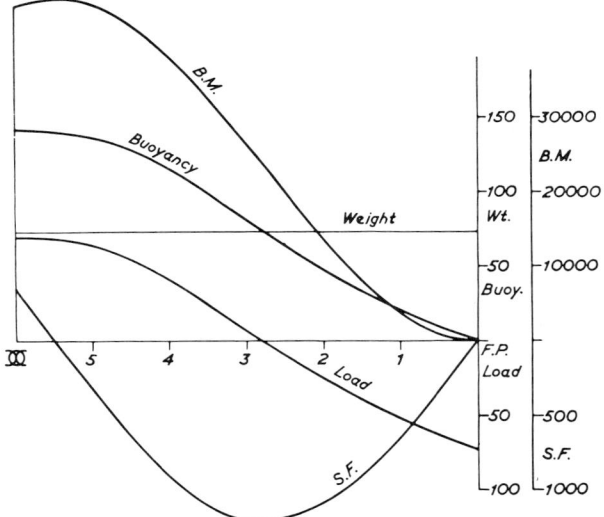

FIG. 14–6—*Example 14.1.*

ALTERNATIVE SOLUTION

Buoyancy Tonnes/metre	SM	f	Lever	f
0	1	—	6	—
21·7	4	86·8	5	434·0
48·3	2	96·6	4	386·4
81·0	4	324·0	3	972·0
113·0	2	226·0	2	452·0
136·5	4	546·0	1	546·0
143·0	1	143·0	0	—
		1,422·4		2,790·4

Buoyancy $= 1{,}422\cdot4 \times \tfrac{1}{3} \times 10\cdot7 = 5{,}075$ tonnes
Weight $= 6 \times 10\cdot7 \times 73 = 4{,}685$ tonnes
 SF at ⊗ $= \overline{390\text{ tonnes}}$

LCB $= \dfrac{2{,}790\cdot4}{1{,}422\cdot4} \times 10\cdot7 = 21\cdot0$ m forward of ⊗

STRENGTH OF SHIPS   227

Moment of weight about $\overline{\omega\omega}$ = 4,685 × 32·1 = 150,500 tonne metre
Moment of buoyancy about $\overline{\omega\omega}$ = 5,075 × 21·0 = 106,500 tonne metre
Bending moment at $\overline{\omega\omega}$ = 44,000 tonne metre
= 431·8 MN metre

EXAMPLE 14.2.
A ship of length 130 metres has the undernoted mean values for weight and buoyancy as measured at the mid-point of each of ten equi-spaced stations. Determine the bending moment at amidships.

Station	$\frac{1}{2}$	$1\frac{1}{2}$	$2\frac{1}{2}$	$3\frac{1}{2}$	$4\frac{1}{2}$	$5\frac{1}{2}$	$6\frac{1}{2}$	$7\frac{1}{2}$	$8\frac{1}{2}$	$9\frac{1}{2}$
Weight	75·5	111·0	126·1	125·3	88·8	106·7	96·8	120·8	124·4	103·9 tonne/metre
Buoyancy	53·3	93·3	115·5	125·3	131·5	133·3	133·3	123·3	105·7	64·9 tonne/metre

SOLUTION

Station	Weight	Buoyancy	Load	SF	SF areas	BM areas
0	—	—	—	0	—	0
$\frac{1}{2}$	75·5	53·3	−22·2	—	−22·2	—
1	—	—	—	−22·2	—	22·2
$1\frac{1}{2}$	111·0	93·3	−17·7	—	−62·1	—
2	—	—	—	−39·9	—	84·3
$2\frac{1}{2}$	126·1	115·5	−10·6	—	−90·4	—
3	—	—	—	−50·5	—	174·7
$3\frac{1}{2}$	125·3	125·3	0	—	−101	—
4	—	—	—	−50·5	—	275·7
$4\frac{1}{2}$	88·8	131·5	42·7	—	−58·3	—
5	—	—	—	−7·8	—	334·0
$5\frac{1}{2}$	106·7	133·3	26·6	—	11·0	—
6	—	—	—	18·8	—	323
$6\frac{1}{2}$	96·8	133·3	36·5	—	74·1	—
7	—	—	—	55·3	—	249
$7\frac{1}{2}$	120·8	123·3	2·5	—	113·1	—
8	—	—	—	57·8	—	136
$8\frac{1}{2}$	124·4	105·7	−18·7	—	96·9	—
9	—	—	—	39	—	39
$9\frac{1}{2}$	103·9	64·9	−39	—	39	—
10	—	—	—	0	—	0

Bending moment at amidships = 334·0 × 13 × 13 × $\frac{1}{2}$
= 28,220 tonnes metre = 276·9 MN metre
Shearing force at amidships = 7·8 × 13
= 102 tonnes = 1·02 MN

Approximations for Maximum Bending Moments and Shearing Forces

It is apparent that a great deal of information is necessary in order to calculate the bending moment and shearing force on a ship. It is desirable to have a means of making an approximation to the maximum values of shearing force and bending moment as this is frequently required before the full information is available.

It is reasonable to relate the maximum bending moment to the product of the displacement and length.

Thus Maximum bending moment $\propto \Delta \times L$

or $$BM = \frac{\Delta L}{C} \; tm \quad \text{or} \quad = \frac{\Delta L}{C} \times 9813 \; Nm.$$

where C is a coefficient and Δ is the total displacement.

The coefficient C depends upon a number of factors—load distribution, type of ship, distribution of buoyancy—and consequently it is difficult to give values of C which will be suitable for general application. It is desirable to determine the values of C for different types of ships at different conditions of loading and tabulate these as data for future reference.

Approximate values for C for specific cases are given in Table 14.3.

Table 14.3.

Ship Type	L (metres)	Δ (Tonnes)	Sagging	C Hogging
Cargo	130	12,600	35	33
Tanker	160	22,500	40	95
Tanker	220	62,500	40	115
Liner	225	41,000	116	30
	290	64,000	80	30

The maximum shearing force may be related to the displacement and in the standard condition of loading is frequently taken as

$$\frac{\Delta}{10} \; \text{tonnes} \quad \text{or} \quad \frac{\Delta}{10} \times 9813 \; N$$

An expression used for the maximum shearing force is

$$SF = \frac{C \cdot BM_{\text{amidships}}}{L} \; \text{tonnes}$$

where the coefficient $C \doteqdot 4$

or $$= \frac{C \cdot BM_{\text{amidships}}}{L} \times 9813 \; N$$

Modulus and Stress Calculations

Having determined the total bending moment (BM) or (M) the stress is calculated by the simple beam theory using the relation

$$\frac{P}{Y} = \frac{M}{I} \quad \text{or} \quad P = \frac{M}{I/Y}$$

where P = stress in MN/m² at distance Y from the neutral axis
M = bending moment at section considered in tonne metres
I = moment of inertia of the section considered about the neutral axis in m² cm²

I/Y will have its smallest value when Y is greatest; that is when Y is measured to the extreme fibres, at the deck and keel. This value of I/Y or Z is called the section modulus and is the criterion of the strength of the girder in bending. If the neutral axis is not at half depth there will be two values of Z i.e. I/Y_1 and I/Y_2 one of which will give the greatest tensile stress and the other the greatest compressive stress.

The determination of the longitudinal stress value for a known bending moment requires the assessment of I and Y for the cross-section of the structure in way of the maximum bending moment. In the calculation for the section modulus it is the accepted practice to take the cross section at amidships and in way of openings. Figure 14-7 is an

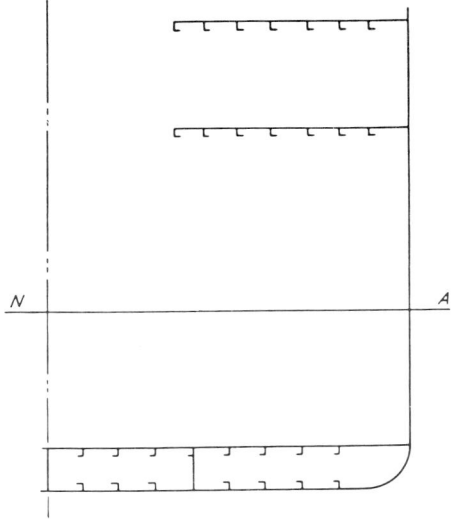

FIG. 14-7—*Midship section of ship—indicating longitudinal strength members.*

outline midship section of a ship showing the longitudinal material which may be taken to be effective in resisting bending moment. Only material which is distributed over a considerable length in the fore and aft direction should be included. Classification societies normally require the continuation of midship scantlings over an appreciable length of the ship before commencing to taper off towards the ends.

The items which are, in general, always included are all continuous decks, deck longitudinals, side and bottom shell plating, bottom longitudinals, tank top plating with margin plate and centre girder. There will be other items depending upon the type of ship such as oil tankers where there would be longitudinal bulkheads. In large passenger ships with superstructures these could be included if they extend over a sufficiently great length.

As the moment of inertia of the cross-section is determined with reference to the neutral axis the position of that axis is an essential requirement in the calculation. It is not necessary to calculate the position of the neutral axis first and then the moment of inertia about this axis. This can be done in one calculation by the use of the principle of parallel axes since

$$I_{NA} = I_{xx} - Ad^2$$

where I_{NA} = moment of inertia about the neutral axis
I_{xx} = moment of inertia about some axis XX
A = total cross sectional area of material
d = distance from XX to neutral axis.

The choice of the XX axis is quite arbitrary but is commonly taken at about 40 to 45 per cent the depth of section above the base. This will bring it reasonably near the expected position of the neutral axis and consequently reduce the magnitude of the figures when calculating moments and moments of inertia.

EXAMPLE 14.3.

The calculation is best carried out in tabular form and the section modulus (I/Y) calculation relating to Figure 14-7 is given in Table 14.4. All items on each side of the assumed (xx) axis should be grouped together under the headings of "above assumed axis" and "below assumed axis" as shown in Table 14.4.

In the first column of Table 14.4 the items are listed and the sectional areas are given in column 2. Only half the area of the double bottom centre girder is taken as this is the only item of the structure which does not appear on both sides of the ship and there will be a multiplication by 2 for both sides. Column 3 gives the distance of the centroid of the individual item from the assumed axis. The product of the area and this distance AY are given in column 4. This is the moment of area about the assumed axis. This product is again multiplied by the lever giving AY^2 and shown in column 5. To secure the moment of inertia of an item about the assumed axis it is essential to add to AY^2 the moment of inertia of the item about its own neutral axis. This is shown in column 6. Most of the items can be assumed as rectangles so

STRENGTH OF SHIPS 231

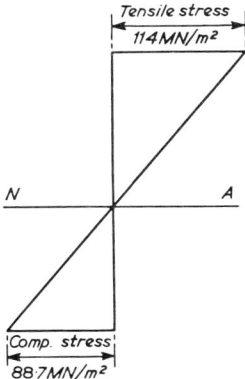

FIG. 14–8—*Longitudinal stress.*

that the moment of inertia is $1/12\, Ah^2$ where A is the area of the item and h the depth. It is only necessary to deal with 'vertical' material as the moment of inertia of 'horizontal' material about its own neutral axis is so small that it may be neglected.

Since the axis has been assumed above the base there is material above and below the axis so that to determine the position of the neutral axis the difference of the moments in column 4 must be taken. In this example the total moment above the assumed neutral axis (19053) is greater than the moment below the assumed neutral axis (13778). Thus the actual neutral axis is above that assumed by an amount d where

$$d = \frac{\text{net moment}}{\text{total area}} = \frac{5275}{7069 \cdot 6} = 0.746 \text{ m.}$$

Table 14.4. Calculation for Section Modulus $[I/Y]$
Assumed neutral axis 4·5 m above keel. Depth to U deck = 12·0 m
Above Assumed Axis

1	2	3	4	5	6
Item	Area $[A]$ $[cm^2]$	Y $[m]$	AY $cm^2\,m$	AY^2 $[m^2\,cm^2]$	$\dfrac{Ah^2}{12}$ $[m^2\,cm^2]$
UD stringer angle	60·0	7·42	445	3,300	—
UD plating	1,210·5	7·73	9,350	72,400	—
UD longitudinals	107·2	7·67	820	6,300	—
2nd D plating	574·3	4·97	2,860	14,200	—
longitudinals	95·4	4·91	468	2,300	—
Sheerstrake	308·2	6·92	2,130	14,750	63 [$h = 1\cdot57$]
Side shell	978·4	3·05	2,980	9,100	3,030 [$h = 6\cdot1$]
Total:					
Above assumed axis	3,334·0		19,053	122,350	3,093

Below Assumed Axis

Side shell	525·0	1·65	866	1,430	470 [$h = 3·28$]
Bilge strake	214·2	3·81	815	3,100	46 [$h = 1·6$]
Bottom shell	1,240·4	4·51	5,590	25,200	—
Keel	147·2	4·57	674	3,080	—
Tank top centre line	90·2	3·42	308	1,050	—
Tank top plating	935·4	3·42	3,200	10,950	—
Tank top longitudinal	190·1	3·54	672	2,380	—
Bottom shell longitudinal	209·5	4·40	924	4,060	—
Centre girder	78·6	4·00	314	1,260	9 [$h = 1·14$]
Side girder	105·0	3·96	415	1,640	10 [$h = 1·09$]
Total:					
Below assumed axis	3,735·6		13,778	54,150	535
Above assumed axis	3,334·0		19,053	122,350	3,093
	7,069·6		5,275	176,500	3,628
				3,628	
				180,128	

Neutral axis (NA) above assumed axis $= \dfrac{5,275}{7,069·6} = 0·746$ m

Neutral axis (NA) above base $= 4·5 + 0·746$
$= 5·246$ m

I about assumed axis (one side) $= 180,128$ m² cm²
I about NA (one side) $= 180,128 - 7,069·6 \times (0·746)^2$
$= 176,200$
I about NA for both sides $= 176,200 \times 2 = 352,400$ m² cm²
Y_1 to upper deck $= 12·0 - 5·246 = 6·754$ m

∴ Section modulus (I/Y_1) to deck $= \dfrac{352,400}{6·754} = 52,180$ m cm²

Y_2 to keel $= 5·246$ m

∴ Section modulus (I/Y_2) to keel $= \dfrac{352,400}{5·245} = 67,170$ m cm²

In the assessment of longitudinal stress values it is the maximum values that are desired so that the maximum bending moment and the maximum value of Y at the section where this takes place are desired. For the ship for which the calculations are shown in Table 14.4 the maximum hogging bending moment was 60,700 tonne metre or 595·5 MN metre. Thus:

Stress in deck $= \dfrac{MY_1}{I} = \dfrac{60,700 \times 6·754}{352,400} = \dfrac{\text{tonne metre} \times \text{metre}}{\text{m}^2 \text{ cm}^2}$
$= 1·163 \times 9,813 \times 100^2/10^6$
$= 114$ MN/m² tension or $\dfrac{595·5 \times 6·754 \times 100^2}{352,400}$
$= 114$ MN/m²

Stress in bottom $= \dfrac{MY_2}{I} = \dfrac{60,700 \times 5·246}{352.400} \times 9,813 \times 100^2/10^6$
$= 88·7$ MN/M² compression

STRENGTH OF SHIPS 233

The stress varies linearly from a maximum tensile value at the outside fibres on one side to a maximum compressive value at the outside fibres on the other side as shown in Figure 14.8.

The maximum allowable stress in terms of the length of the ship has been variously estimated and two examples of these modified for metric units are as follows:

1) $P = 77 + 0.25\,L$ where $P = $ MN/m² and L is in metres

L	60	100	140	180	220	260	300 m
P	92	102	112	122	132	142	152 MN/m²

2) $P = 23\,\sqrt[3]{L}$

	90	106	119	129	138	146	154 MN/m²

Shear Stress

The shearing force at any position of the ships length is that force which tends to move one part of the ship relative to the adjacent portion. The shearing force is generally at its maximum at about a quarter of the length from each end. The maximum shearing stresses occur in the neighbourhood of the neutral axis.

The shear stress at any point is given by

$$q = \frac{FA\bar{Y}}{bI}$$

where $q = $ shear stress at any point
$F = $ the shearing force at the section under consideration
$A\bar{Y} = $ moment of area about the NA above or below the surface under consideration
$I = $ total moment of inertia about the NA
$b = $ total thickness of material resisting shear

EXAMPLE 14.4.

In a ship the maximum shearing force is 44·76 MN. $A\bar{Y}$ about the neutral axis is 151,000 m cm². I about the neutral axis is 2,758,000 m² cm² and the thickness of the shell plating at the neutral axis is 2·14 cm. Determine the shear stress at the neutral axis.

$q = \dfrac{FA\bar{Y}}{2bI}$ Note: 2 is required for both sides of the ship.

$= \dfrac{44\cdot76 \times 151{,}000}{2 \times 2\cdot14 \times 2{,}758{,}000} = \dfrac{\text{MN} \times \text{m cm}^2}{\text{cm} \times \text{m}^2\,\text{cm}^2} = \dfrac{\text{MN}}{\text{cm} \times \text{m}}$

$= \dfrac{44\cdot76 \times 151{,}000 \times 100}{2 \times 2\cdot14 \times 2{,}758{,}000}$

$= 57\cdot24$ MN/m²

EXAMPLE 14.5.

In a calculation for the modulus of section of a ship having a depth of

12·5 m the axis is assumed at 5 m above the keel and the following results were obtained for one side only:

	A	AY	AY^2	$Ah^2/12$
	cm^2	$cm^2\ m$	$cm^2\ m^2$	$cm^2\ m^2$
Above axis	3,160	14,850	84,800	240
Below axis	3,260	12,050	49,700	180

Determine the modulus of the section and the maximum direct stresses in the section for a hogging bending moment of 44,500 tonne metre or 436·7 MN m:

	A	AY	AY^2	$1/12\ Ah^2$
Above axis	3,160	14,850	84,800	240
Below axis	3,260	12,050	49 700	180
	6,420	2,800	134,500	420
			420	
			134,920	

NA above assumed axis $= \dfrac{2,800}{6,420} = 0·436$ m

$\qquad\qquad\qquad\qquad\quad = 5 + 0·436 = 5·436$ m above keel

I about assumed axis $\quad = 134,920$ cm² m²

I about NA $\qquad\qquad\quad = 134,920 - 6,420 \times (0·436)^2$

$\qquad\qquad\qquad\qquad\quad = 133,700 \times 2 = 267,400$ cm² m² (both sides)

Modulus of section $= I/Y \qquad\qquad Y$ to deck $= 12·5 - 5·436$
$\qquad\qquad\qquad\qquad\qquad\qquad\qquad\qquad\quad = 7·064$ m
$\qquad\qquad\qquad = \dfrac{267,400}{7·06}$

$\qquad\qquad\qquad = 37,800$ cm² m

From $P/Y = M/I$: P at deck $= \dfrac{44,500 \times 7·064}{267,400} = \dfrac{\text{tonne metre} \times \text{metre}}{cm^2\ m^2}$

$\qquad\qquad\qquad\qquad\quad = 1·175 \times 9,813 \times 100^2/10^6$

$\qquad\qquad\qquad\qquad\quad = 115·3$ MN/m²

$\qquad\qquad P$ at keel $= \dfrac{44,500 \times 5·436}{267,400}$

$\qquad\qquad\qquad\qquad\quad = 0·905 \times 9,813 \times 100^2/10^6$

$\qquad\qquad\qquad\qquad\quad = 88·8$ MN/m²

EXAMPLE 14.6.

The midship section of a ship of breadth 16·5 m and depth 11 m can be assumed as shown in Figure 14–9. All the material has a thickness of 1·25 cm.

STRENGTH OF SHIPS 235

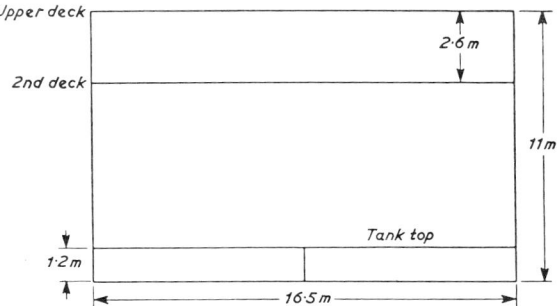

FIG. 14-9—*Example 14.6.*

Determine the moment of inertia of the section about the neutral axis. In this example the moments are taken about the base.

Item	Area (cm)²	Lever m	Moment	Lever m	I	I/12 ah²
Upper deck	2,062	11·0	22,682	11·0	249,500	
2nd deck	2,062	8·4	17,320	8·4	145,500	
Tank top	2,062	1·2	2,474	1·2	2,970	
Bottom shell	2,062	—	—	—	—	
Side shell	2,750	5·5	15,130	5·5	83,220	27,730
Cr. girder	150	0·6	90	0·6	54	18
	11,148		57,696		481,244	27,748
					27,748	
					508,992	

$$\text{Neutral axis above base} = \frac{57{,}696}{11{,}148}$$
$$= 5 \cdot 18 \text{ m}$$

I about base $= 508{,}992 \text{ cm}^2 \text{ m}^2$

I about NA $= I$ about base $- Ad^2$
$= 508{,}992 - 11{,}148 \times 5 \cdot 18^2$
$= 209{,}890 \text{ cm}^2 \text{ m}^2$

EXAMPLE 14.7.
The midship section of a ship may be assumed as rectangular with a breadth of 18 m and depth 9 m. There is a second deck 2·5 m below the upper deck and a double bottom 1·25 m deep with a centre girder. The material throughout is 1·53 cm thick with the exception of the second deck and tank top which are 1·27 cm thick. Determine the stress in the upper deck when the ship is subjected to a sagging bending moment of 31,000 tonne metre or 304·2 MN m. If the limiting compressive stress in the deck is taken as:

$$\frac{278}{1 + \frac{1}{950}\left[\frac{s}{t}\right]^{1.75}}$$

s = beam spacing
t = thickness of deck plating

determine the maximum spacing of the upper deck beams.

Item	Area cm²	Lever m	Moment	Lever m	I	1/12 Ah²
Upper deck	2,754	9	24,786	9	223,070	
2nd deck	2,286	6·5	14,860	6·5	96,580	
Tank top	2,286	1·25	2,857	1·25	3,570	
Bottom shell	2,754	—	—	—	—	
Sides	2,754	4·5	12,392	4·5	55,770	18,590
Cr. girder	191	0·62	118	0·62	70	25
	13,025		55,013		79,060	18,615
					318,615	
					397,675	

$$\text{NA above base} = \frac{55,013}{13,025}$$
$$= 4\cdot 22 \text{ m}$$
I about base $= 397,675$
I about NA $= 397,675 - 13,025 \times 4\cdot 22^2$
$= 165,670 \text{ cm}^2 \text{ m}^2$

From $P = \dfrac{MY}{I}$

Y to deck $= 9 - 4\cdot 22$
$= 4\cdot 78 \text{ m}$

or

$$= \frac{31,000 \times 4\cdot 78}{165,670} = 0\cdot 894$$

$$\frac{304\cdot 2 \times 4\cdot 78 \times 100^2}{165,670}$$

$= \dfrac{\text{tonne metre} \times \text{metre}}{\text{cm}^2 \text{ m}^2}$

$= 87\cdot 7 \text{ MN/m}^2$

$= 0\cdot 894 \times 9\cdot 813 \times 100^2/10^6$

$= 87\cdot 7 \text{ MN/m}^2$

$$87\cdot 7 = \frac{278}{1 + \dfrac{1}{950}\left[\dfrac{s}{1\cdot 53}\right]^{1.75}}$$

$$87\cdot 7 + \frac{87\cdot 7}{950}\left[\frac{s}{1\cdot 53}\right]^{1.75} = 278$$

$$\left[\frac{s}{1\cdot 53}\right]^{1.75} = \frac{278 - 87\cdot 7}{0\cdot 0923} = \frac{190\cdot 3}{0\cdot 0923} = 2,062$$

$s = 120 \text{ cm}$

Stresses in the Inclined Condition

In the standard longitudinal strength calculation it is assumed that the ship is upright and that the bending moment is in the vertical plane. This being so the ship bends about a horizontal axis and consequently the moment of inertia is calculated about such axis. This does not necessarily give the greatest stress in the structure. When the ship is inclined the depth of the section is increased and increased stresses could be experienced at the corners.

Consider the ship inclined at some angle θ to the vertical as indicated in Figure 14–10. The bending moment will still be in the

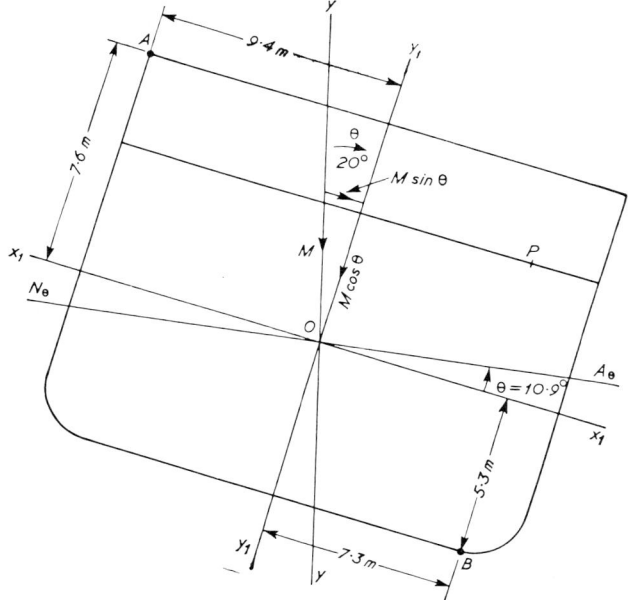

FIG. 14–10—*Stresses in the inclined condition.*

vertical plane and the problem is one of unsymmetrical bending. Let the bending moment in the vertical plane be M; this moment can be resolved into two moments, one in the plane of y called M_y and one in the plane of x called M_x.

Thus
$$M_y = M \cos \theta \quad \text{and} \quad M_x = M \sin \theta$$

If x and y are the co-ordinates of any point (P) in the structure and I_{NA} and I_{CL} are the moments of inertia about the horizontal axis in the upright condition and about the centre line respectively then the stress at P with co-ordinates xy is given by

Stress at P.
$$P = \frac{M \cos \theta \, y}{I_{NA}} + \frac{M \sin \theta \, x}{I_{CL}}$$

At the neutral axis corresponding to the inclined position the stress is zero hence the position of the neutral axis is determined by the condition that

$$\frac{\cos\theta\ y}{I_{NA}} + \frac{\sin\theta\ x}{I_{CL}} = 0$$

or

$$y = -\frac{I_{NA}}{I_{CL}}\tan\theta\ x$$

This gives the equation to the neutral axis in the inclined condition. This is inclined to the neutral axis in the upright condition at an angle φ given by

$$\tan\varphi = \frac{y}{x} = -\frac{I_{NA}}{I_{CL}}\tan\theta$$

The neutral axis as inclined ($N_\theta\ A_\theta$) is a straight line through 0 making an angle φ with the neutral axis ($X_1\ X_1$) in the vertical condition. In Figure 14–10 the ship is inclined through the angle θ in the clockwise direction and the angle φ is measured from NA in the counter-clockwise direction.

In the event that $I_{NA} = I_{CL}$ then $\tan\varphi = -\tan\theta$ and the neutral axis is horizontal. In general this is unlikely as I_{CL} is normally about twice I_{NA} and consequently the neutral axis is inclined to the horizontal.

The angles at which the greatest and least stresses occur will be given by putting $dp/d\theta = 0$

$$\frac{dp}{d\theta} = -\frac{M\sin\theta y}{I_{NA}} + \frac{M\cos\theta x}{I_{CL}} = 0$$

or

$$\tan\theta = \frac{x}{y}\frac{I_{NA}}{I_{CL}}$$

The greatest and least stresses will also be associated with the maximum values of x and y and thus these stresses will occur at the corners of the section.

EXAMPLE 14.8.

In Figure 14–10 the points most distant from the neutral axis are A and B with the co-ordinates in metres of A with respect to the axis $X_1 X_1$ and $Y_1 Y_1$ (9·4, 7·6) and for B (7·3, 5·3). For the section shown the I_{NA} is $0·348 \times 10^6$ cm^2 m^2 and I_{CL} is $0·655 \times 10^6$ cm^2 m^2.

For an inclination of 20 degrees

$$\tan\varphi = -\frac{0·348 \times 10^6}{0·655 \times 10^6} \times 0·3640$$

$$= -0·1933 \quad \therefore\ \varphi = -10·9 \text{ degrees}$$

STRENGTH OF SHIPS

For a hogging bending moment of 53,000 tonne metre or 520 MN metre in the upright condition:

at point A

$$\text{stress } P = M \left[\frac{\cos \theta y}{I_{NA}} + \frac{\sin \theta x}{I_{CL}} \right]$$

$$= 53,000 \left[\frac{0.9397 \times 7.6}{0.348 \times 10^6} + \frac{0.3420 \times 9.4}{0.655 \times 10^6} \right]$$

$$= 1.088 + 0.260 = 1.348 \; \frac{\text{tonne metre} \times \text{metre}}{\text{cm}^2 \, \text{m}^2}$$

$$= 1.348 \times 9.813 \times 100^2/10^6$$

$$= 132 \text{ MN/m}^2 \text{ tension}$$

or $520 \, [2.055 \times 10^{-5} + 0.49 \times 10^{-5}]$

$= 520 \times 2.54 \times 10^{-5} \times 10^4$

$= 132 \text{ MN/m}^2$

at point B

$$P = 53,000 \left[\frac{0.9397 \times 5.3}{0.348 \times 10^6} + \frac{0.3420 \times 7.3}{0.655 \times 10^6} \right]$$

$$= 96.0 \text{ MN/m}^2 \text{ compression.}$$

Modifying the Strength Section

Increase in depth of section does not necessarily reduce the maximum stress. The value of I will be increased but also the value of Y. If the increase in Y is relatively greater than the increase in I, the ship's girder is not strengthened by the addition of material.

If it is assumed that a suitable standard of strength for a given bending moment is that the stress at the highest deck should not exceed that at the top of the main structure then

$$\frac{dy}{y} = \frac{dI}{I}$$

For example assume that a (cm)^2 of deck plating are added at a height h (m) above the main structure of sectional area A (cm^2) and moment of inertia I (cm^2 m^2). Let Y be distance from original neutral axis to top of main structure and let $f = Y + h$. Then

$$\text{Shift of neutral axis} = \frac{af}{A + a}$$

New I about original axis $= I + af^2$

New I about new axis $= I + af^2 - [A + a] \left[\frac{af}{A + a} \right]^2$

$$= I + \frac{Aaf^2}{A + a}$$

To maintain standard of strength the stress in the highest deck must not exceed that at top of main structure and the condition is that

$$\frac{I}{Y} = \frac{I_1}{Y_1}$$ See Figure 14-11 $I_1 = MI$ of new section

$Y_1 = $ distance from new NA to added deck

$$= I\frac{[A + a] + Aaf^2}{Af}$$

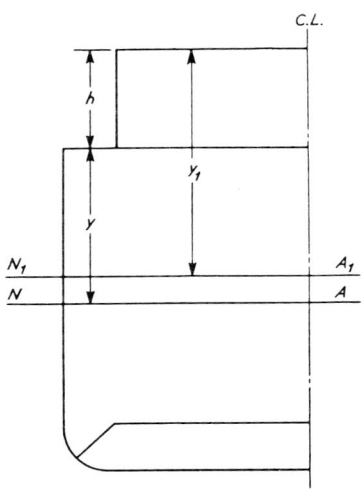

FIG. 14-11—*Modification of strength section.*

thus
$$IA + Ia + Aaf^2 = \frac{IAf}{Y}$$

and
$$aI + Aaf^2 = \frac{IAh}{Y}$$

$$a = \frac{I}{Y}\frac{Ah}{Af^2 + I}$$

EXAMPLE 14.9.

A superstructure deck of breadth 9·25 m is to be fitted 2·75 m above the top of the ship's main structure. Determine the thickness of this deck plating so that the stress in the superstructure deck will not exceed that at the top of the main structure without the superstructure.

$A = 22{,}600$ cm²; $INA = 268{,}000$ cm² m²; NA to top of main hull $= 7{\cdot}6$ m

Let $t = $ thickness of superstructure deck plating in cm. Then:

STRENGTH OF SHIPS

$$9 \cdot 25 \times 100 \times t = \frac{268{,}000}{7 \cdot 6} \cdot \frac{22{,}600 \times 2 \cdot 75}{22{,}600 \times 10 \cdot 35^2 + 268{,}000}$$

$$t = \frac{268{,}000 \times 22{,}600 \times 2 \cdot 75}{2{,}688{,}500 \times 7 \cdot 6 \times 925}$$

$$= 0 \cdot 88 \text{ cm}$$

EXAMPLE 14.10.

For the conditions of Example 14.8 determine (1) the rise in the neutral axis due to the superstructure deck and (2) the new I about the new axis.

$$\text{Rise in NA} = \frac{af}{A+a} = \frac{9 \cdot 25 \times 100 \times 0 \cdot 88 \times 10 \cdot 35}{22{,}600 + 814}$$

$$= \frac{8{,}423}{23{,}414} = 0 \cdot 36 \text{ m}$$

New I about new axis $= I + \dfrac{Aaf^2}{A+a}$ $\quad f = 7 \cdot 6 + 2 \cdot 75$
$\quad = 10 \cdot 35$ m

$$= 268{,}000 + \frac{22{,}600 \times 814 \times 10 \cdot 35^2}{23{,}414}$$

$$= 268{,}000 + 84{,}160 = 352{,}160 \text{ cm}^2 \text{ m}^2$$

or

$$\frac{I}{Y} = \frac{I_1}{Y_1} \qquad Y_1 = 7 \cdot 6 - 0 \cdot 36 + 2 \cdot 75 = 9 \cdot 99 \text{ m}$$

$$\frac{268{,}000}{7 \cdot 6} = \frac{I_1}{9 \cdot 99}$$

$$\therefore \qquad I_1 = \frac{268{,}000 \times 9 \cdot 99}{7 \cdot 6} = 352\,160 \text{ cm}^2 \text{ m}^2$$

Deflexion of Ships

The longitudinal deflexion of a ship is mainly due to a change of bending moment.

It is assumed that a ship behaves as a variably loaded and variably supported girder and that stresses and deflexions can be assessed by means of the simple beam theory.

The three quantities, curvature $1/R$ slope and deflexion are interconnected in the same way as load, shearing force and bending moment. Although $1/R = M/EI$ it is generally more convenient to prepare a curve of M/I and to divide the derived deflexion by E.

That is the deflexion Y caused by bending is given by:

$$Y = \frac{1}{E} \int \int \frac{M}{I} \, dx \, dx$$

where

$Y =$ deflexion (when adjusted) in cm
$M =$ bending moment in tonne metre
$I =$ moment of inertia in cm^2 m^2
$E =$ modulus in MN/cm^2

With a beam for simple cases of loading the value M is obtained by a simple mathematical expression and if the beam is of constant section then I is constant and the integration is a straightforward calculation.

For a ship the calculation of the deflexion of the structure when subjected to longitudinal bending is not quite so straightforward. This is because the bending moment on the structure cannot be represented by a simple expression and again the value of I is not constant along the length of the ship. Consequently it is necessary to determine the value of the moment of inertia at a number of positions throughout the ship's length and then plot a curve of M/I.

The integration of the curve of M/I can be carried out graphically or as is usual in tabular form by dividing the length of the ship into a number of sections and taking the mean ordinate for each division.

This first integration of M/I will give a curve of slope and the second integration carried out in the same way will give a curve of deflexion. The ordinates of such a curve may be termed the total deflexions between the ends of the vessel but since the deflexion is of necessity measured with reference to a straight line joining the ends, the final deflexion curve is obtained by measuring between such a line and the second integral curve and setting up these distances on a horizontal base line.

The procedure is illustrated by the following example.

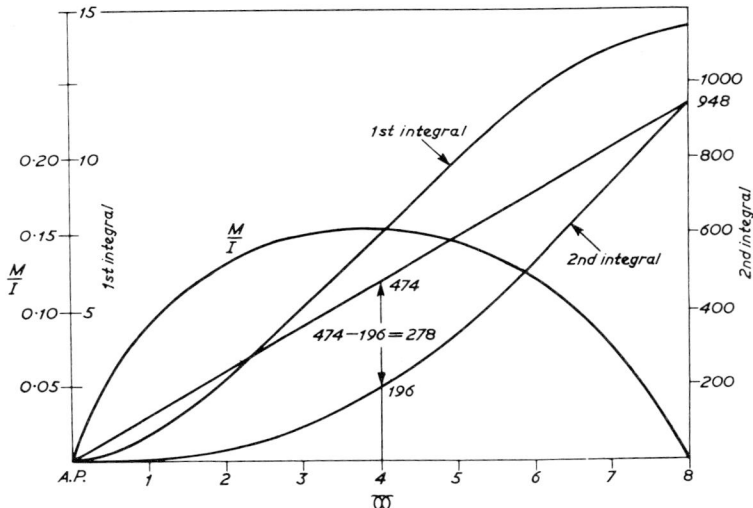

FIG. 14-12--*Example 14.11.*

EXAMPLE 14.11.

Estimate the deflexion of the hull of a ship amidships relative to the straight line through its extremities at the perpendiculars when the values of M/I (tonne metre/cm² m²) at equidistant stations commencing at the after perpendicular are as follows:

STRENGTH OF SHIPS 243

0, 0·093, 0·135, 0·150, 0·156, 0·145, 0·119, 0·078, 0

The length of the ship is 128 m. Assume $E = 20.9$ MN/cm^2.

From Figure 14–12 and taking mean ordinates.

Common interval $= 128/8 = 16$ metres

	1st Integral			2nd Integral	
To Station 1	$16 \times 0.057 =$	0·912	To Station 1	$16 \times 0.25 =$	4
1–2	$\times 0.115 =$	1·84	1–2	$\times 1.75 =$	28
	To 2 $=$	2·75		To 2 $=$	32
2–3	$\times 0.144 =$	2·30	2–3	$\times 4.0 =$	64
	To 3 $=$	5·05		To 3 $=$	96
3–4	$\times 0.155 =$	2·48	3–4	$\times 6.25 =$	100
	To 4 $=$	7·53		To 4 $=$	196
4–5	$\times 0.152 =$	2·43	4–5	$\times 8.75 =$	140
	To 5 $=$	9·96		To 5 $=$	336
5–6	$\times 0.135 =$	2·16	5–6	$\times 11.0 =$	176
	To 6 $=$	12·12		To 6 $=$	512
6–7	$\times 0.10 =$	1·6	6–7	$\times 13.0 =$	208
	To 7 $=$	13·72		To 7 $=$	720
7–8	$\times 0.044 =$	0·70	7–8	$\times 14.2 =$	228
	To 8 $=$	14·42		To 8 $=$	948

$$\text{Deflexion} = \frac{1}{E} \int \int \frac{M}{I} \, dx \, dx$$

$$\text{Units of } Y = \frac{1}{\text{MN/cm}^2} \cdot \frac{\text{tonne metre}}{\text{cm}^2 \, \text{m}^2} \, \text{m.m.}$$

$$\text{Deflexion} = \frac{278}{20.9} \times \frac{9{,}813}{10^6} \times 100 = 13 \text{ cm}$$

EXAMPLE 14.12.

The mean values for weight and buoyancy as measured at centres between each of six displacement stations are as follows:

Station	Weight (Tonnes/m)	Buoyancy (Tonnes/m)
$\frac{1}{2}$	43·51	17·99
$1\frac{1}{2}$	85·62	69·64
$2\frac{1}{2}$	47·98	85·96
$3\frac{1}{2}$	41·31	73·97
$4\frac{1}{2}$	34·65	46·98
$5\frac{1}{2}$	52·11	10·66

The moment of inertia of section is assumed constant at 60,000 cm^2 m^2. Determine the deflexion due to bending at amidships. The length of the ship is 78 metres, E can be taken as 20·9 MN/cm^2.

Station	Weight	Buoy-ancy	Load	SF	SF areas	M areas	BM	M/I
0				0		0	0	
½	43·51	17·99	25·52		25·52			
1				25·52		25·52	2,156	0·0359
1½	85·62	69·64	15·98		67·02			
2				41·50		92·54	7,820	0·1303
2½	47·98	85·96	−37·98		45·02			
3				3·52		137·56	11,620	0·194
3½	41·31	73·97	−32·66		−25·62			
4				−29·14		111·94	9,458	0·158
4½	34·65	46·98	−12·33		−70·61			
5				−41·47		41·33	3,492	0·0582
5½	52·13	10·66	41·47		−41·47			
6				0		0	0	

For bending moment multiply M areas by ½ × 13 × 13, i.e. 84·5. See Figure 14-13.

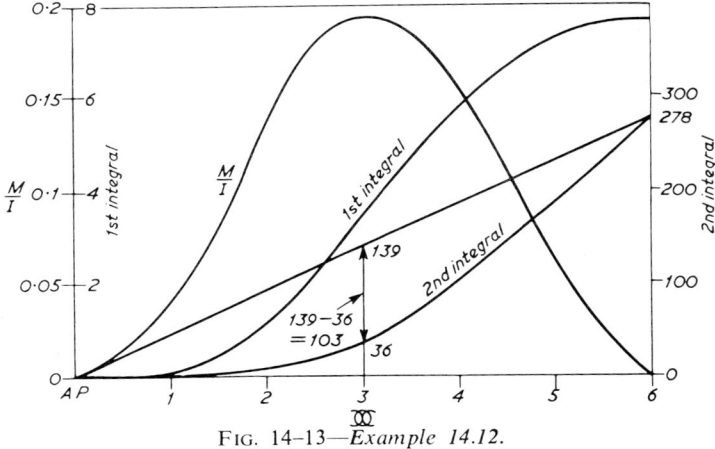

FIG. 14-13—*Example 14.12.*

	1st Integral	CI = 13 m	2nd Integral	
To Station 1	13 × 0·015 = 0·19	To Station 1	13 × 0·08 =	1·04
1–2	× 0·077 = 1·00	1–2	× 0·50 =	6·5
	To 2 = 1·19		To 2 =	7·54
2–3	× 0·175 = 2·28	2–3	× 2·2 =	28·6
	To 3 = 3·47		To 3 =	36·14
3–4	× 0·185 = 2·40	3–4	× 4·7 =	61·2
	To 4 = 5·87		To 4 =	97·34
4–5	× 0·105 = 1·36	4–5	× 6·6 =	84·5
	To 5 = 7·23		To 5 =	181·84
5–6	× 0·02 = 0·26	5–6	× 7·5 =	95·6
	To 6 = 7·49		To 6 =	277·44

$$\text{Deflexion} = \frac{1}{E} \int \int \frac{M}{I} \, dx \, dx$$

$$\text{Units of } Y = \frac{1}{\text{MN/cm}^2} \frac{\text{tonne metre}}{\text{cm}^2 \text{ m}^2} \text{ m.m.}$$

$$\text{Deflexion} = \frac{103}{20 \cdot 9} \times \frac{9{,}813}{10^6} \times 100 = 4 \cdot 8 \text{ cm}$$

Experiments on Longitudinal Strength of Ships

Experimental work on the longitudinal strength of ships can be divided into three categories:
   (a) Full scale static experiments;
   (b) Full scale experiments at sea;
   (c) Experiments on special models in a towing tank.

(*a*)   *Static Experiments*

Due to the loss of two naval vessels at sea in 1901 and the finding of a court martial that the loss was due to structural weakness of the ships a committee was set up to investigate the strength of the ship type involved. The experiments on the destroyer HMS *Wolf* were the first comprehensive experiments on the strength of ships.

The *Wolf* was floated into a dry dock and the water gradually pumped out. As the water level fell the ship settled on two specially constructed cradles and it was possible from the water levels and the hydrostatic data of the ship to calculate the bending moment to which the ship was subjected. Hogging and sagging conditions were considered by arranging the cradles in different positions. Structural strains were measured by means of mechanical strain gauges and deflexions of the structure by means of posts erected in the dock. Due to the difficulty of measuring strain at that time the results are now largely of historical interest.

A number of full scale static tests have been carried out over the years including tests on two British tankers. These were initiated by the Admiralty Ship Welding Committee. Two similar tankers were chosen one the *Neverita* which was mainly welded and the other the *Newcombia* which was mainly riveted.

Changes in bending moment were applied to the ship, which was afloat in a river, by means of shifting water ballast. Strains were measured by means of mechanical, electrical resistance and acoustic gauges and measurements were also made of local deflexions of plating.

From the evidence of all the experiments carried out it can be accepted that:
   (1) Where the structure is continuous over an appreciable part of the length the longitudinal bending stress can be computed with sufficient accuracy from simple beam theory;

(2) The formula for shear stress $q = FA\overline{Y}/bI$ may be used to assess the greatest shear stress in the structure;
(3) A superstructure is effective in making a contribution to the strength of the ship.

(*b*) *Experiments at Sea*

A number of tests have been carried out on ships at sea but none so complete as the trials on the *Ocean Vulcan* and the *Clan Alpine*. The instrumentation on the ship consisted of normal pressure indicators on the hull surface, wave profile indicators, accelerometers capable of measuring the accelerations in three directions, gyroscopic roll and pitch recorders giving angle of yaw, and strain gauges from which plate strain could be deduced. In addition there were wind speed and direction indicators. A survey of wave heights was made by means of stereo cameras.

Details of these experiments are given in Report No. 8 of the Admiralty Ship Welding Committee.

Since the experiments on the *Ocean Vulcan* and *Clan Alpine* were carried out a great deal of data has been collected at sea from other ships. These observations tend to confirm that the standard calculation of the stress due to waves is satisfactory as a means of comparison between ship and ship, and between various loading conditions for the same ship.

(*c*) *Experiments on Special Models*

The models used for investigating bending moments amongst waves are very different from those used for resistance experiments. They are of two types. In the first type the model is constructed of metal and strain gauges are used to measure the strain at a particular section. The second type of model is hinged and consists of two parts which are attached by means of a hinge and metal strips.

The procedure adopted with these models is to run them head-on into waves of predetermined height and length and to obtain records of the strain experienced at a particular section and also of the ship motions.

A great deal of experimental work has been done on ship models in waves and the method of investigation is extending the knowledge of the strength of ships amongst waves.

Pressure on Immersed Area

The mean pressure on any immersed surface is that due to the head at the centre of area or centroid of the surface.

This mean pressure multiplied by the area gives the total pressure.

So that $P = HAw$

where P = total pressure in Newton:

H = depth of centroid of area from surface of liquid in metres
A = immersed area in m²
w = specific weight of liquid in N/m³
= density of liquid in kg/m³ × 9·81.

EXAMPLE 14.13.

A rectangular double bottom tank is 18 metres long and 12 metres in breadth. The tank is filled with sea water until the water is 9 metres up the overflow pipe above the tank top. Determine the pressure in N/m² and the total pressure on the tank top.

$P = HAw$ 　　　　Pressure on tank top = 18 × 12 × 90,500
= 9 × 1 × 1,025 × 9·81　　　　　　　　　= 19·55 × 10⁶ N
= 90,500 N/m²　　　　　　　　　　　　　= 19·55 MN.

EXAMPLE 14.14.

Determine the pressure on the ends and sides of a tank of triangular cross-section with vertex down having a length of 3 m, width 2 m and 1·23 m deep when filled with water at 1,000 kg/m³. Figure 14–14.

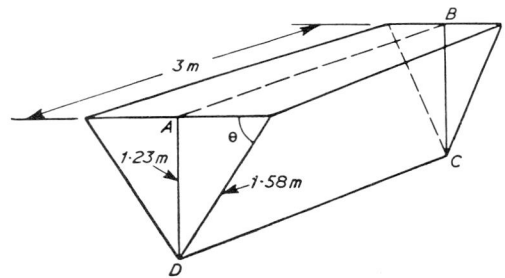

FIG. 14–14—*Example 14.14.*

1) Pressure on one end = Depth to centroid × A × W
$$= \frac{1·23}{3} \times \tfrac{1}{2} \times 2 \times 1·23 \times 10^3 \times 9·81$$
= 4,950 N

2) Pressure on side = Depth to centroid × A × w
$$= \frac{1·23}{2} \times 3 \times 1·58 \times 10^3 \times 9·81$$
= 28,650 N (P_n)

3) Vertical component of pressure on side = $P_n \cos \theta$　　$\cos \theta = \dfrac{1}{1·58}$

$$= 28,650 \times \frac{1}{1·58} = 18,130 \text{ N}$$

Thus vertical pressure for both sides = 2 × 18,130 = 36,260 N which is the weight of water in the tank, i.e. ½ × 2 × 1·23 × 3 × 10³ × 9·81 = 36,260 N

4) Horizontal pressure on one side $= P_n \sin \theta$

$$= 28{,}650 \times \frac{1 \cdot 23}{1 \cdot 58} = 22{,}300 \text{ N}$$

This is the same as the pressure on the vertical centreline longitudinal plane ($ABCD$)

Pressure on plane $ABCD$ = Depth to centroid $\times A \times W$
$$= 0 \cdot 615 \times 3 \times 1 \cdot 23 \times 10^3 \times 9 \cdot 81$$
$$= 22{,}300 \text{ N}$$

Centre of Pressure

The point through which the resultant of the total pressure acts is known as the centre of pressure. It is at the centroid of the load diagram.

Since the pressure at any point of an immersed surface depends upon the depth of the point below the fluid surface it follows that the pressure is uniform along horizontal lines and variable along vertical lines.

The distribution of the pressure on the vertical plane can be shown by a rectangular and vertical plane as in Figure 14–15 where the

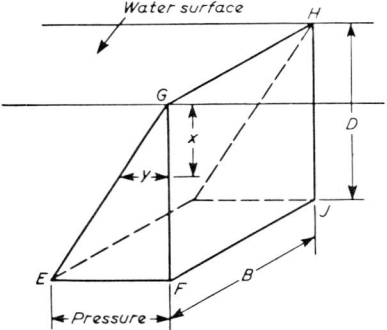

FIG. 14–15—*Distribution of pressure on vertical plane.*

breadth $= FJ$; depth $= HJ$; the upper edge GH is at the fluid surface.

On any vertical line such as GF the pressure varies from zero at G to $1{,}025 \times 9 \cdot 81$ D newtons at F. If this pressure intensity at F is set off to some convenient scale such as FE then the triangle GFE becomes a pressure diagram and the horizontal intercept y at any depth x below the fluid surface gives the pressure at that depth. The mean pressure is that due to the mean head $D/2$ at the mid position of GF.

The resultant of the pressure acts through the centroid of the pressure diagram, that is, at $2/3$ D below the fluid surface.

STRENGTH OF SHIPS

EXAMPLE 14.15.

Consider a bulkhead rectangular in form with a width of 6 metres, depth 8 m and loaded on one side to the top edge with sea water. Then:

Load/m at top of bulkhead $= 0$, since the head is zero

Load/m at bottom of bulkhead $=$ head $\times$ width $\times$ density
$= 8 \times 6 \times 1{,}025 \times 9 \cdot 81 = 482{,}000$ N

Pressure on bulkhead $=$ area of pressure diagram. Figure 14-16
$= \frac{1}{2} \times 482{,}000 \times 8$
$= 1{,}928{,}000$ N $= 1 \cdot 928$ MN

Centre of pressure $=$ centroid of pressure diagram
$= \frac{1}{3} \times 8 = 2 \cdot 66$ m above base.

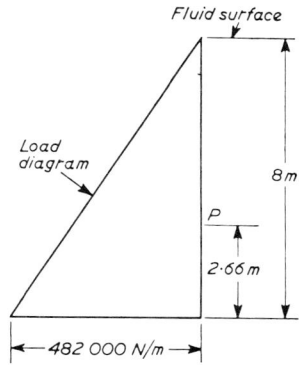

FIG. 14–16—*Example 14.15.*

Position of Centre of Pressure (CP)

Distance of CP from fluid surface

$$= \frac{I \text{ of immersed area about fluid surface}}{\text{Distance of centroid of immersed area from fluid surface} \times \text{immersed area}}$$

$$= \frac{\text{2nd moment of area about surface}}{\text{1st moment of area about surface}}$$

The moment of inertia of a plane area about a given axis and the Theorem of Parallel Axes have already been considered.

EXAMPLE 14.16.

(a) Determine the depth of the centre of pressure of a rectangular plate, in the vertical plane, with one edge in the surface of the fluid. Fig. 14-17.

I about $SS = \frac{1}{3} BD^3$

Moment of area about $SS = BD \times D/2 = \dfrac{BD^2}{2}$

∴ CP below fluid surface $= \frac{1}{3} BD^3 \div \dfrac{BD^2}{2} = \frac{2}{3} D$

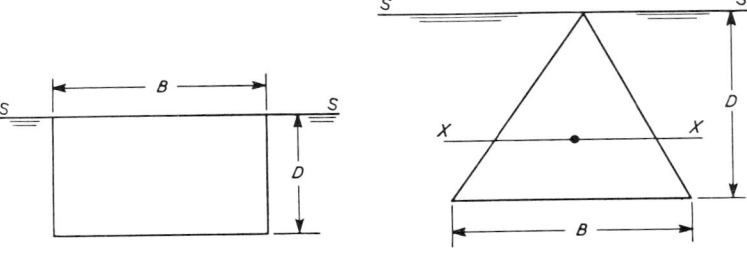

Fig. 14–17—*Example 14.16.* Fig. 14–18—*Example 14.16.*

(b) Triangular plate in the vertical plane with its apex in the surface of the fluid. Figure 14-18

$$I \text{ about } XX = \frac{BD^3}{36} \qquad Y = \text{distance between } SS \text{ and } XX$$

$$I \text{ about } SS = \frac{BD^3}{36} + Ay^2$$

$$= \frac{BD^3}{36} + \left[\frac{BD}{2} \times \frac{4}{9} D^2\right] = \frac{BD^3}{4}$$

∴ $\quad CP \text{ below fluid surface} = \frac{BD^3}{4} \div \frac{BD^2}{3} = \tfrac{3}{4} D$

Values of I, centroid of area and centre of pressure for certain geometrical shapes are given in Table 14.5.

Table 14.5.

Depth of Centre of Pressure

Figure	Depth of centroid below surface	I for axis at centroid	Depth of centre of pressure below surface
Rectangle depth D, breadth B upper edge at fluid surface	$D/2$	$1/12\ BD^3$	$\tfrac{2}{3} D$
Triangle height D, base B apex at fluid surface	$\tfrac{2}{3} D$	$1/36\ BD^3$	$\tfrac{3}{4} D$
Triangle height D, base B edge in surface	$\tfrac{1}{3} D$	$1/36\ BD^3$	$\tfrac{1}{2} D$
Circle diameter D with circumference touching surface	$D/2$	$\dfrac{\pi D^4}{64}$	$\tfrac{5}{8} D$

STRENGTH OF SHIPS 251

If top edge is below the fluid surface these values undergo an appreciable change as shown in the following example.

EXAMPLE 14.17.

An isosceles triangle is immersed as shown in Figure 14–19. The triangle has a width at top of 7·5 m and a depth of 7 m. Determine the position of the centre of pressure when the fluid surface is 4 m above top edge.

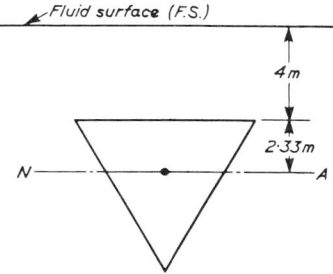

FIG. 14–19—*Example 14.17.*

$I_{FS} = I_{NA} + Ay^2$ $Y = \tfrac{1}{3} \times 7 + 4 = 6\cdot33$ m
$= 1/36\ BD^3 + Ay^2$
$= 1/36 \times 7\cdot5 \times 7^3 + \tfrac{1}{2} \times 7\cdot5 \times 7 \times 6\cdot33^2$
$= 715 + 1{,}052$
$= 1{,}767$ m² m²

$CP = \dfrac{1{,}767}{166}$ 1st moment $= \tfrac{1}{2} \times 7\cdot5 \times 7 \times 6\cdot33$
 $= 166$

$= 10\cdot64$ m below fluid surface

Centre of Pressure on Inclined Plane

The rule for vertical planes

CP from fluid surface $= \dfrac{I \text{ about fluid surface}}{\text{Moment of area about surface}}$

say $Hp = \dfrac{I_o}{Ax_a}$

can be modified for inclined planes.
From Figure 14–20:

$x_p = \dfrac{I_o}{Ax_a}$ and $x_a = \dfrac{Ha}{\sin \theta}$

$Hp = x_p \sin \theta = \dfrac{I_o \sin \theta}{Ax_a}$

$= \dfrac{I_o \sin \theta \sin \theta}{AHa} = \dfrac{I_o \sin^2 \theta}{AHa}$

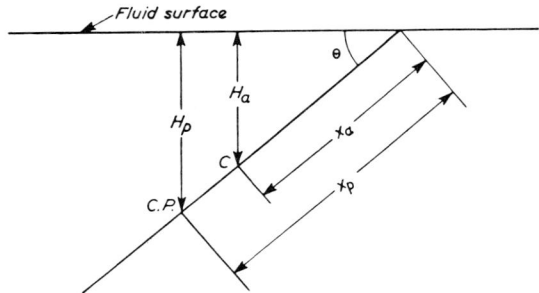

FIG. 14-20—*Centre of pressure on inclined plane.*

Watertight Bulkheads

A watertight bulkhead is in effect a stiffened sheet of plating restrained all round its edges and subjected to water pressure on one side. Normally the bulkhead stiffeners are vertical and the problem is that of the bending of a simple beam with a regularly increasing load.

A bulkhead stiffener supports a rectangle of plating the area of which is equal to the length of the stiffener (l) multiplied by the spacing of stiffeners (s). If the bulkhead is flooded on one side to the top edge, the stiffener supports a pressure which increases uniformly from zero at the top to a maximum at the bottom as shown in Figure 14–21 (a) and (b).

Pressure on Bulkhead and Centre of Pressure

EXAMPLE 14.18.

The widths of a bulkhead commencing at the top and at intervals of 1·5 m are as follows:

6·1, 5·5, 4·6, 3·6 and 2·4 metres.

The bulkhead is flooded on one side to the top with sea water. Determine the area of and the pressure on the bulkhead together with the centre of pressure.

Width	SM	f	Lever	f of 1st moment	Lever	f of 2nd moment
6·1	1	6·1	0	—	0	—
5·5	4	22·0	1	22·0	1	22·0
4·6	2	9·2	2	18·4	2	36·8
3·6	4	14·4	3	43·2	3	129·6
2·4	1	2·4	4	9·6	4	38·4
		54·1		93·2		226·8

STRENGTH OF SHIPS 253

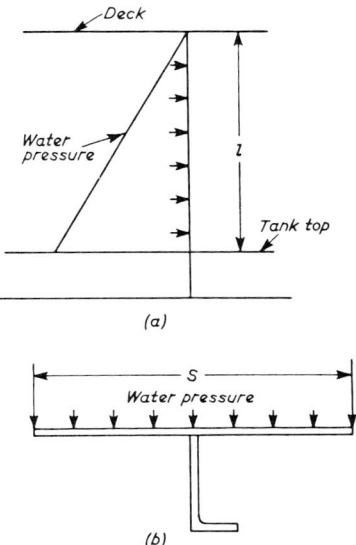

FIG. 14–21—*Watertight bulkheads.*

Area of bulkhead $= 54 \cdot 1 \times \tfrac{1}{3} \times 1 \cdot 5$
$\phantom{\text{Area of bulkhead }}= 27 \cdot 05 \text{ m}^2$
Pressure on bulkhead $=$ density $\times 9 \cdot 81$ 1st moment of area about fluid surface
$\phantom{\text{Pressure on bulkhead }}= 1{,}025 \times 9 \cdot 81 \times 93 \cdot 2 \times \tfrac{1}{3} \times 1 \cdot 5 \times 1 \cdot 5$
$\phantom{\text{Pressure on bulkhead }}= 703{,}000 \text{ N}$

or

Centroid of area below top of bulkhead
$$= \frac{93 \cdot 2}{54 \cdot 1} \times 1 \cdot 5 = 2 \cdot 58 \text{ m}$$

Pressure on bulkhead $= H \, A \, w$
$\phantom{\text{Pressure on bulkhead }}= 2 \cdot 58 \times 27 \cdot 05 \times 1{,}025 \times 9 \cdot 81$
$\phantom{\text{Pressure on bulkhead }}= 703{,}000 \text{ N as above.}$

Centre of pressure from fluid surface $= \dfrac{\text{2nd moment of area}}{\text{1st moment of area}}$

$$= \frac{226 \cdot 8}{93 \cdot 2} \times 1 \cdot 5$$
$$= 3 \cdot 65 \text{ m}$$

EXAMPLE 14.19.

A bulkhead is flooded on one side to the top edge with sea water. The vertical stiffeners have a length of 9 m, spaced 0·75 m apart and free at each end. Determine the total load on a stiffener, the shearing force at the ends, the position of zero shear and the maximum bending moment.

Pressure at x metres from top
$$= x \times 0.75 \times 1,025 \times 9.81$$
$$= 7,540\ x\ \text{N/m}$$
$$\text{Total pressure} = \int_0^9 7,540\ xdx = 3,770 \times 9^2$$
$$= 305,400\ \text{N} = 30.54 \times 10^4\ \text{N}$$
Taking moments about bottom of bulkhead
$$R_T \times 9 = 30.54 \times 10^4 \times \tfrac{1}{3} \times 9$$
$$\therefore R_T = 10.18 \times 10^4\ \text{N and}$$
$$R_B = 20.36 \times 10^4\ \text{N}$$

The shearing force at a distance x from top of bulkhead is the reaction at the top less the pressure at x m from top.
$$SF \text{ at depth } x = R_T - \int_0^x 7540\ x\ dx$$
$$= 10.18 \times 10^4 - 3770\ x^2$$
so that when $x = 0$ SF at top $= 10.18 \times 10^4$ N
$x = 9$ SF at bottom $= 20.36 \times 10^4$ N

when SF is zero the bending moment is at a maxmium
$$SF = 0 \text{ when } 10.18 \times 10^4 = 3770\ x^2$$
$$\therefore x = 5.20 \text{ metres}$$
$$BM \text{ maximum} = \int_0^{5.20} 10.18 \times 10^4 - 3770\ x^2\ dx$$
$$= \left[10.18 \times 10^4\ x - 3770 \times \tfrac{1}{3} x^3 \right]_0^{5.20}$$
$$= 10.18 \times 10^4 \times 5.2 - 1257 \times 5.20^3$$
$$= 53.0 \times 10^4 - 17.7 \times 10^4$$
$$= 35.3 \times 10^4\ \text{Nm}$$

Figure 14–22 shows a bulkhead flooded on one side up to the bulkhead deck. Consider the part of the bulkhead from the 2nd deck to the tank top. There is then in this case a head of sea water above the top of the stiffeners which extend from the tank top to the 2nd deck.

EXAMPLE 14.20.

For the bulkhead shown in Figure 14–22 for the portion between the tank top and the 2nd deck the stiffeners are spaced 0·75 m apart. Draw the pressure, shearing force and bending moment diagram for a stiffener which is free at the ends. State the maximum value of BM.

Pressure at depth x metres from water surface $= x \times 0.75 \times 1,025 \times 9.81$
$$= 7,540\ x\ \text{N/m}$$
Pressure at 2nd deck $= 7,540 \times 2.5 = 18,850$ N/m
Pressure at tank top $= 7,540 \times 7.5 = 56,550$ N/m

STRENGTH OF SHIPS

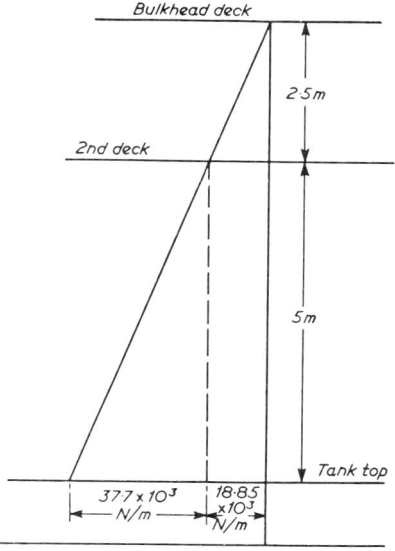

FIG. 14-22—*Bulkhead flooded on one side.*

∴ Total load = $\frac{1}{2}$ (18·85 × 10³ + 56·55 × 10³) × 5 = 188·5 × 10³ N
Load diagram may be divided into a rectangle and a triangle as in Figure 14-22.

Taking moments about tank top:
R_T × 5 = 18·85 × 10³ × 5 × 2·5 + 0·5 × 5 × 37·7 × 10³ × 5/3
= 235·6 × 10³ + 157·1 × 10³ = 392·7 × 10³

∴ $R_T = \dfrac{392·7 \times 10^3}{5} = 78·54 \times 10^3$ N

R_B = 109·96 × 10³ N

SF at x metres from 2nd deck

$= R_T - \int_0^x 7{,}540 \,(2·5 + x)\, dx$

$= R_T - \int_0^x 18·85 \times 10^3 + 7{,}540\, x\, dx$

= 78·54 × 10³ − (18·85 × 10³ × x + $\frac{1}{2}$ × 7,540 x^2)
= 78·54 × 10³ − 18·85 × 10³ x − 3,770 x^2 N

thus $x^2 + 5x - 20·83 = 0$
$(x + 7·7)(x - 2·7) = 0$

∴ SF zero at x = 2·7 metres

x from 2nd deck	0	1·25	2·50	3·75	5·0 metres
SF	78·5	49·1	7·8	−44·8	−109·9 × 10³ N

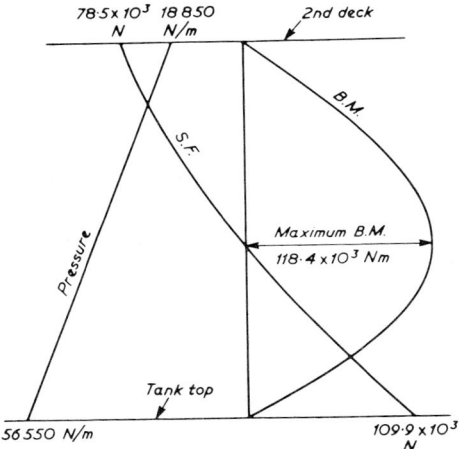

Fig. 14–23—*Example 14.20.*

BM at x metres from 2nd deck

$$= \int_0^x 78\cdot54 \times 10^3 - 18\cdot85 \times 10^3 x - 3{,}770\, x^2$$

$= 78\cdot54 \times 10^3\, x - 9\cdot42 \times 10^3\, x^2 - 1{,}257\, x^3$ Nm

x from 2nd deck	0	1·25	2·50	3·75	5·0	metres
BM	0	81·0	117·7	95·7	0	$\times 10^3$ Nm

BM at maximum when SF is zero.

∴ maximum $BM = 78\cdot54 \times 10^3 \times 2\cdot7 - 9\cdot42 \times 10^3 \times 2\cdot7^2 - 1{,}257 \times 2\cdot7^3$

$\quad = 212 \times 10^3 - 68\cdot8 \times 10^3 - 24\cdot8 \times 10^3$

$\quad = 118\cdot4 \times 10^3$ Nm

CHAPTER 15

Vibration of Ships

Vibration is a subject of considerable importance to naval architects and marine engineers because of its adverse effects both upon the structure of the ship and upon the comfort of passengers and crew. Due to the complexity of the ship's structure and the numerous possible sources of vibration that exist on a ship the complete solution of all problems in ship hull vibration is not yet available.

The vibration experienced on ships may be divided broadly into two classes.

1) *Synchronous or Resonant Vibration*

In this the entire hull girder is put into a state of vibration at certain revolutions of the main propelling machinery or of some auxiliary machinery. Clearly if such vibration occurs at revolutions in the region of those required in service the position is serious as it is quite impracticable to run a ship in such circumstances.

2) *Local Vibration*

In this, isolated parts of the ship or certain fittings such as panels of plating are set into a state of vibration. Vibrations of this nature are not dangerous to the ship but can be very disturbing to passengers and crew. In general the cure is to eliminate, if possible, the source of the trouble. Failing this, local treatment in the form of additional pillars and stiffeners becomes necessary.

It must be appreciated that once a ship is a completed unit it is virtually impossible to eliminate resonant vibration by adding stiffening material and consequently attention must be directed to the source of the disturbing force.

The golden rule to avoid vibration is to remove its cause; unfortunately this is not always possible.

The hull of a ship will only vibrate if some external force is applied to it. Continuous vibration that leads to structural damage and the discomfort of personnel on board is due to out-of-balance periodic forces in the propelling machinery, shafting, propellers or auxiliary machinery.

A ship when afloat may be regarded as an elastic beam or structure without rigid constraint and consequently is susceptible to vibration. The features of ship vibration can be illustrated from a consideration of the ways in which a homogeneous bar of constant section vibrates. A bar of this type supported at its ends could vibrate as shown in Figure 15-1 between contours ABC and ADC. In this case

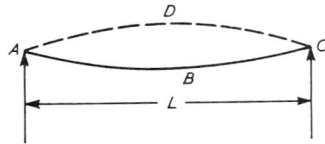

FIG. 15-1—*Vibration of a homogeneous bar of constant section.*

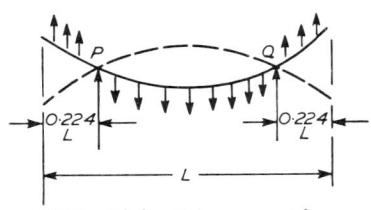

FIG. 15-2—*Primary mode.*

the centre of gravity of the bar would oscillate in space. However, the naval architect is interested in the type of vibration in which the centre of gravity does not move. This requires that the ends of the bar should be moving upwards at the same time as the middle of the bar is moving downwards. The bar then vibrates as a free-free beam and will do so if the supports are 0·224 L from the ends. This is shown in Figure 15-2 and is called the *Primary Mode* for it has the least number of nodes; a node is a point of rest; the points P and Q do not move.

The time of one vibration, from one extreme position to the other extreme position and back again is known as the *Period*. The *Frequency* is the number of vibrations which take place in unit time and is the inverse of the period.

The frequency of a homogeneous constant section free-free beam is given by

$$N = \frac{n^2}{2\pi} \sqrt{\frac{EIg}{WL^3}}$$

where E = modulus of elasticity of the material of the beam
I = the moment of inertia of its section about the neutral axis
W = weight of the bar
L = length of the bar
N = a numerical constant that depends on the mode of vibration.

Empirical formulae for the frequency of vertical vibration of a ship are, in general, an adaptation of the formula given above; a change is made in the coefficient.

The hull of a ship can be considered as an elastic girder of box-like construction, in which the distribution of weight and of the moment of inertia of the cross-section of the structural material vary from one end to the other. As such the hull possesses natural frequencies of different modes vertically, laterally and about a longitudinal torsional axis.

The possibilities of vibration from one cause or another are high and consequently it is essential to examine the entire problem in the design stage. If the probable natural frequencies of the hull can be assessed within reasonable limits then attention can be given to the choice of the revolutions per minute of the propelling machinery, the number of propeller blades and the balancing of all such units.

It is very desirable to have means of estimating the critical frequencies of the hull in the early stages of a ship design. Once the ship is built little can be done to alter the natural frequencies.

The first systematic investigation of ship-hull vibration was made by Otto Schlick in 1884.

The empirical formula proposed by Schlick was as under

$$\text{Frequency} = \phi \sqrt{\frac{I}{\Delta L^3}}$$

where ϕ is a coefficient best calculated from data for a ship similar to the one under consideration.

Todd suggests the following values of ϕ Column (1) for the two node vertical vibration.

	(1)	(2)	(3)
Large tankers fully loaded	130,000	282,000	28,000
Small tankers fully loaded	100,000	217,000	21,500
Cargo ships at about 60 per cent of load displacement	112,000	243,000	24,100

where imperial units are used, and N = frequency per minute; I = moment of inertia in in^2/ft^2 of all longitudinally continuous material in the cross-section amidships; Δ = displacement of ship in tons; L = length of ship in feet.

If I is in m^4; Δ in MN and L is in m then the constants have values as shown in Column (2).

If I is in $\text{m}^2 \text{ cm}^2$; Δ in tonnes and L in m then the constants have values as shown in Column (3).

A defect in the Schlick formula as originally stated is that it disregards the effect of the virtual added weight due to the surrounding water. The total mass which is being vibrated equals the displacement of the ship plus a mass of water. The displacement of the ship may thus be considered to be multiplied by a factor called the virtual inertia factor, to give the total virtual mass.

It has been suggested by Todd that:
$$\text{Virtual inertia factor} = \frac{B}{3H} + 1\cdot 2$$
where B = breadth of ship and H = draught

then
$$N = \phi \sqrt{\frac{I}{\Delta_1 L^3}}$$

where to take account of the entrained water, a virtual weight Δ_1 is used instead of the displacement Δ

and
$$\Delta_1 = \Delta \left(\frac{B}{3H} + 1\cdot 2\right)$$

so that:
$$N = \phi \sqrt{\frac{I}{(B/3H + 1\cdot 2)\Delta L^3}}$$

Another expression is that given by Burrill. This is an extension of the Schlick formula based on the work of Taylor. Burrill concluded that the total virtual weight Δ_1 could be expressed by the equation
$$\Delta_1 = \Delta \left(1 + \frac{B}{2H}\right)$$

He also added a factor $\sqrt{1+r}$ to allow for sheer correction on the basis of Taylor's work and gave a chart of $\sqrt{1+r}$ for different values of B/D and D/L which is reproduced in Figure 15-3.

The final expression is:
$$N = \frac{\phi}{\sqrt{(1 + B/2H)(1+r)}} \sqrt{\frac{I}{\Delta L^3}}$$

where imperial units are used and:
B = breadth of ship in feet; H = draught in feet;
I = moment of inertia in in^2 ft^2;
Δ = displacement of ship in tons; L = length of ship in feet;
D = depth in feet.

For these units ϕ is given as 200,000.

$$r = \frac{3\cdot 5 D^2 (3a^3 + 9a^2 + 6a + 1\cdot 2)}{L^2 (3a + 1)} \quad \text{where } a = \frac{B}{D}$$

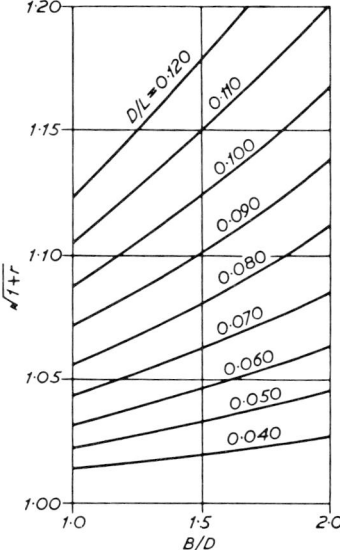

Fig. 15-3—Values of $\sqrt{1+r}$ for different values of B/D.

If I is in m² cm²; Δ in tonnes; L, B, D and H in metres then ϕ has a value of 44,000.

A formula frequently used is based on a suggestion made by Liddell and is as follows:

$$N = 110,000 \sqrt{\frac{I}{\Delta_1 L^3}} + 29$$

where N = 2 node vertical frequency per min;
I = moment of inertia in in² ft²;
L = length of ship in ft;

$$\Delta_1 = \Delta \left(1 \cdot 2 + \frac{B}{3H}\right) \text{ tons}$$

where breadth (B) and draught (H) are in ft.

If I is in m² cm²; Δ_1 in tonnes, B, H and L in metres then the constant is 24,000.

The Schlick formula and those derived from it involve the moment of inertia I of the midship section and this is not always available during the early stages of a design. This is overcome in the Todd formula which is similar to Schlick's but using only the principal dimensions of the ship.

The expression is

$$N = \beta \sqrt{\frac{BD^3}{\Delta L^3}}$$

where imperial units are used and B = breadth of ship in feet; D = depth of ship to strength deck in feet, Δ = displacement in tons; L = length of ship in feet.

The values of β are given below in column (1)

	(1)	(2)	(3)
Large tankers fully loaded	61,000	11,000	112,000
Small tankers fully loaded	45,000	8,150	82,500
Cargo ships at about 60 per cent of load displacement	51,000	9,200	93,500

If I is in m⁴; Δ in MN and L, B and D in metres then the constants are as in column (2).

If I is in m² cm²; Δ in tonnes and L, B and D in metres then the constants have values as shown in column (3).

As in the case of the Schlick formula an improvement is to substitute the total virtual weight, Δ_1, for the displacement, Δ, so that:

$$N = \beta \sqrt{\frac{BD^3}{\Delta_1 L^3}}$$

where, as previously,

$$\Delta_1 = \Delta \left(\frac{B}{3H} + 1 \cdot 2 \right)$$

EXAMPLE 15.1.

A tanker of length 128 m, breadth 16·75 m and load draught 7·34 m has a block coefficient of 0·72. The I is 190,500 m² cm². Determine the natural frequency of vibration. The Schlick constant can be taken as 28,000.

Displacement = 128 × 16·75 × 7·34 × 0·72 × 1·025 = 11,600 tonnes

By Schlick:

$$N = \varphi \sqrt{\frac{I}{\Delta L^3}}$$

$$= \varphi \sqrt{\frac{190,500}{11,600 \times 128^3}}$$

$$= 28,000 \times 0 \cdot 0028$$

$$= 79 \text{ vibrations/minute.}$$

VIBRATION OF SHIPS

EXAMPLE 15.2.

A single deck ship has a length of 122 m, breadth 16·75 m, depth 10·95 m and floats at a uniform draught of 7·32 m with a block coefficient of 0·70. The midship section of the ship can be assumed rectangular and the vessel has a double bottom 1·07 m in depth with a centre girder. All the material can be assumed as 1·25 cm thick. The Schlick constant is 28,000. Determine the natural frequency of vertical vibration.

ITEM	Sectional Area $(cm)^2$	Lever (m)	Moment	Lever (m)	I	$1/12\, Ah^2$
Deck	2,100	10·95	23,000	10·95	252,000	—
Tank top	2,100	1·07	2,240	1·07	2,400	—
Bottom	2,100	—	—	—	—	—
Sides	2,740	5·48	15,050	5·48	82,500	27,400
Cr. girder	134	0·53	71	0·53	38	13
	9,174		40,361		336,938	27,413
					27,413	
					364,351	

$$\text{Neutral axis above base} = \frac{40{,}361}{9{,}174} = 4\cdot4 \text{ m}$$

I about base $= 364{,}351$ m^2 cm^2

I about $NA = 364{,}351 - 9{,}174 \times 4\cdot4^2$

$\phantom{I \text{ about } NA} = 187{,}351$ m^2 cm^2

By Schlick:

$$N = \varphi \sqrt{\frac{I}{\Delta L^3}}$$

$$= 28{,}000 \sqrt{\frac{187{,}351}{10{,}700 \times 122^3}}$$

$= 28{,}000 \times 0\cdot00313$

$= 88$ vibrations/minute

$\Delta = 122 \times 16\cdot75 \times 7\cdot32 \times 0\cdot70 \times 1\cdot025$

$ = 10{,}700$ tonnes.

EXAMPLE 15.3.

A ship of length 134 m, breadth 19·7 m and draught 7·58 m has a displacement of 15,600 tonnes. The I is 364,000 m^2 cm^2. Determine the natural frequency of vibration allowing for entrained water. The constant can be taken as 24,000.

From the expression:

$$N = \varphi \sqrt{\frac{I}{\Delta_1 L^3}} + 29$$

$$= 24{,}000 \sqrt{\frac{364{,}000}{32{,}300 \times 134^3}} + 29$$

$$= 24{,}000 \times 0\cdot00216 + 29$$

$$= 52 + 29$$

$$= 81 \text{ vibrations/minute}$$

where $\Delta_1 = \Delta(1\cdot2 + B/3H)$

$\varphi = 24{,}000$

$$\Delta_1 = 15{,}600 \left(1\cdot2 + \frac{19\cdot7}{22\cdot74}\right)$$

$$= 15{,}600 \times 2\cdot07$$

$$= 32\,300 \text{ tonnes.}$$

CHAPTER 16

Rudders : Oscillations

Rudders

All ships must possess some means of directional control and in most cases this control is effected by means of a rudder located at the after end of the ship. Rudders are fitted at the after end of the ship since in this position they are most effective in giving control and also they derive benefit from the increased water velocity induced by the propellers. This is particularly important at low speed.

An important feature in directional control is what is known as the Neutral Point. This is the point in the length of the ship at which an applied force does not cause the ship to deviate from a constant direction. This neutral point is in general about one-sixth of the length of the ship abaft the bow. Thus if a force is applied abaft the neutral point and acts towards port the ship will turn to starboard; again if applied forward of the neutral point and acts towards port the ship turns to port. Hence, the greater the distance the applied force is from the neutral point the greater will be the turning effect. Thus rudders are more effective when located aft.

The radial force acting on a ship during a steady turn is given by the expression

$$F = \frac{\Delta V^2}{Rg}$$

where F = Newtons; Δ = displacement in tonnes; R = radius of turning circle in m; g = gravity accn. = 9.807 m/sec².

EXAMPLE 16.1.

Determine the radial force required for a ship of 10,000 tonnes displacement turning in a circle of 1,000 m diameter at a steady speed of 16 knots.

$$\text{From } F = \frac{\Delta V^2}{Rg} = \frac{10,000 \, (16 \times 0.5144)^2}{500 \times 9.807}$$

$$= 138.5 \text{ tonnes or } 1.36 \text{ MN.}$$

Angle of Heel when Turning

When a ship is turning steadily in a circle a centrifugal force *(P)* acts at the centre of gravity. The action of the force is to tend to move the ship sideways through the water but this is prevented by the lateral resistance, equal to the centrifugal force, which acts at half-draught ($H/2$). These forces form a heeling couple as shown in Figure 16–1 so that

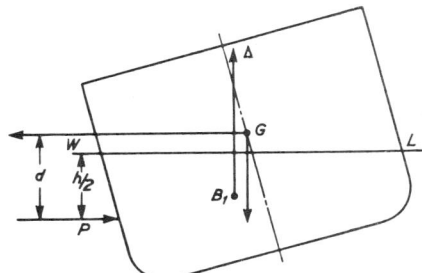

FIG. 16–1—*Heeling couple.*

$$\text{Heeling moment} = \frac{\Delta}{g} \frac{V^2}{R} d$$

where Δ = displacement in tonnes
V = speed in knots
d = distance between forces in metres
g = acceleration due to gravity = 9·807 m/sec²
R = radius of turning circle in metres $\doteqdot$ 3 × length of ship

For small angles of inclination
Righting moment = heeling moment

$$\Delta \times GZ = \Delta\, GM \sin \theta = \frac{\Delta}{g} \times \frac{V^2}{R} \times d$$

$$\therefore \sin \theta = \frac{V^2 d}{g.R.GM}$$

$$= \frac{(V \times 0\cdot5144)^2 \times d}{9\cdot807 \times R \times GM}$$

$$= 0\cdot027\, \frac{V^2 d}{R.GM}$$

where V is in knots; d, R, GM in metres.

EXAMPLE 16.2.

A ship with a metacentric height of 0·91 m has a speed of 12 knots. The centre of gravity is 8·85 m above the keel and the centre of lateral resistance is 3·65 m above the keel. Determine the angle of heel when the rudder is hard over and the ship turns in a circle of diameter 730 m.

From $\sin \theta = 0\cdot027\, \dfrac{12^2 \times 5\cdot2}{365 \times 0\cdot91}$ $\qquad d = 8\cdot85 - 3\cdot65 = 5\cdot2$ m

$\qquad\qquad = 0\cdot061$
$\qquad \therefore \theta = 3\cdot5°$

EXAMPLE 16.3.

A ship turns in a radius of 300 m at a speed of 20 knots under the action of the rudder. If the draught of the ship is 5 m; KG is 6 m and GM_T is 2 m; estimate the approximate angle of heel during the steady turn.

$$\text{From } \sin \theta = 0.027 \, \frac{V^2 d}{R \times GM}$$

$$= 0.027 \times \frac{20^2 \times 3.5}{300 \times 2}$$

$$= 0.063 \quad \therefore \theta = 3.6°$$

$$KG = 6.0 \text{ m}$$
$$H/2 = 2.5 \text{ m}$$
$$d = 3.5 \text{ m}$$

Rudder Types

There are many types of rudder in use by ships today but apart from special rudders and manoeuvring devices they can, broadly, be reduced to four types. The choice of rudder type depends upon the shape of the stern, the area deemed necessary and the capacity of the steering gear to be installed.

The four types are as shown in Figure 16–2.

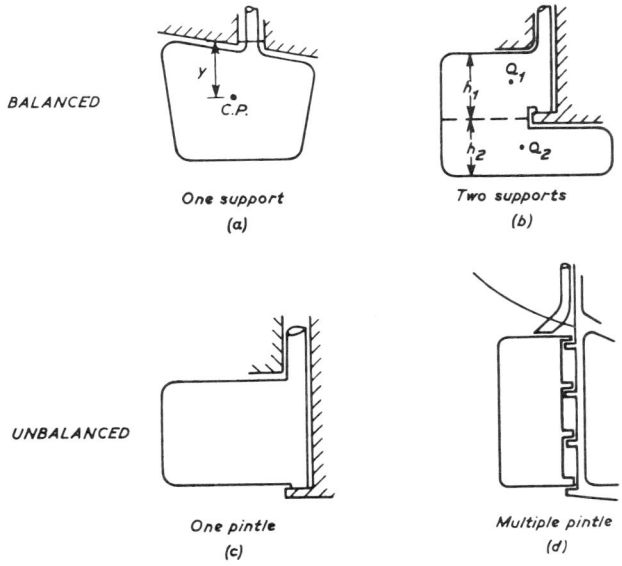

FIG. 16–2—(*a*), (*b*), (*c*), (*d*)—*Rudder types.*

Balanced type (a) one support
type (b) two supports
Unbalanced type (c) one pintle
type (d) multiple pintle

When the total area of the rudder is abaft the axis of rotation the rudder is of the unbalanced type. In the semi-balanced type a proportion of the area is placed forward of the axis of turning.

The balanced type (a) generally known as the spade rudder is adopted where there is considerable cut-up of the stern and where the size of the rudder is not too great.

Balanced type (b) is used where the size of the rudder requires support in addition to that at the bearing and where partial balance is desired to reduce the size of the steering gear.

Unbalanced rudders are adopted when the stern shape is not suitable for the balanced type. The number of pintles used is decided by considerations of strength.

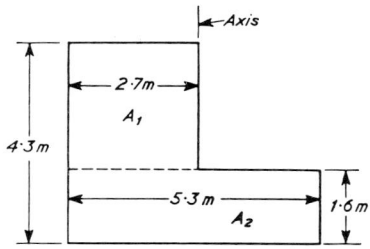

FIG. 16-3—*Centre line rudder on twin screw ship.*

Area

The required area of rudder varies with different types of vessels since the desired manoeuvring ability differs considerably and the general ship design may impose restrictions. In practice the rudder area is usually related to the area of the immersed middle plane. Values of this ratio are normally between 60 and 70. Thus

$$\text{Rudder area} = \frac{L \times H}{60 \text{ to } 70}$$

where $L = LBP$ in metres; $H =$ mean load draught in metres.

The influence of rudder outline apart from aspect ratio—depth/width—is not very great. Thus most ships have rudders tending towards a rectangular outline and as such are simple and efficient. The upper edge of the rudder is generally submerged in the deep load condition in order to reduce the risk of damage by waves in a heavy sea; the upper edge usually follows the contour of the stern.

The current tendency in rudders is towards the double-plate balanced or semi-balanced type of rudder.

Rudder Force

In order to assess the stresses and torque acting on a rudder it is necessary to determine the force on the rudder and its point of application.

The force on the rudder depends upon
(1) the area of the rudder
(2) the form of the rudder
(3) the speed of the ship
(4) the angle of helm. This is limited to 35 degrees to remain outside the stalled condition.

Many investigators have examined the variations of the normal force Q acting on a rudder as the angle of helm (θ) is varied.

Baker and Bottomley suggested for middle line rudders behind single screws a formula which when adjusted for metric units takes the form.

$$Q = 18 \cdot 0 \ A \ V^2 \ \theta \text{ Newtons}$$

where A = rudder area in m²; V = m/sec; θ = degrees

Gawn suggested formulae which when adjusted for metric units are as follows:

For twin rudders behind wing propellers:

Ahead Motion $Q = 21 \cdot 1 \ AV^2 \ \theta$ Newtons
Astern Motion $Q = 19 \cdot 1 \ AV^2 \ \theta$ Newtons

For middle line rudders behind twin screws, both ahead and astern motion:

$$Q = 15 \cdot 5 \ AV^2 \ \theta \text{ Newtons}$$

In these formulae V is taken as the true speed of the ship, allowance is made in the coefficient for propeller race effects.

EXAMPLE 16.4.

A rudder on the middle-line behind twin screws has an area of 13·9 m². The ship speed is 15 knots. Determine the normal force on the rudder at an angle of helm of 35 degrees.

From $Q = 15 \cdot 5 \ A \ V^2 \ \theta$
$= 15 \cdot 5 \times 13 \cdot 9 \times (15 \times 0 \cdot 5144)^2 \times 35$
$= 448,900$ Newtons
$= 448 \cdot 9 \ KN.$

Centre of Pressure

In addition to the normal force on the rudder it is essential to know the position of the centre of pressure. This is the point on the plane at which the resultant force on it may be taken to act. It is necessary to know the torque acting on the rudder to ensure that the steering gear for the ship will be capable of turning the rudder at all speeds.

The French naval constructor Joessell suggested for a flat plate, an empirical formula for the proportion of the breadth of the plate that the centre of pressure is abaft the leading edge; it is expressed as

$$0\cdot195 + 0\cdot305 \sin \theta$$

So that for a rectangular rudder of breadth b—the distance (x) of the centre of pressure from the leading edge is given by

$$x = (0\cdot195 + 0\cdot305 \sin \theta) \, b$$

The leading edge is the fore edge when the ship is going ahead and the aft edge when the ship is going astern. For balanced rudders of rectangular form the centres are measured from the leading edge of the rudder, not from the axis. In general about 25 per cent of the total rudder area is forward of the turning axis.

Gawn suggested that for a rectangular rudder, the centre of pressure is 0·35 times the breadth of the rudder abaft the leading edge if behind deadwood. Figure 16–2 (d) and 0·31 times abaft if the rudder is in the open Figure 16–2 (a). For astern motion the value of 0·31 is used in both cases, the leading edge then being the after edge.

For the semi-balanced rudder Figure 16–2 (b) the normal force and the centre of pressure are calculated by treating the upper part as behind deadwood and the lower part as in the open and combining the separate results. In this case for the centre of pressure the upper part would be subject to the 0·35 rule and the lower part to the 0·31 rule.

Single plate rudders in association with rectangular-section sternframes which were almost universal a few decades ago have been replaced by castings or welded fabrications of streamlined form.

A number of different designs have been produced with variations in detail and some of these in addition to Figure 16–2 are indicated in Figure 16–4.

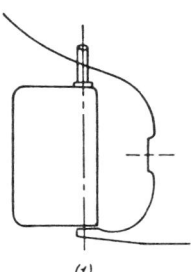 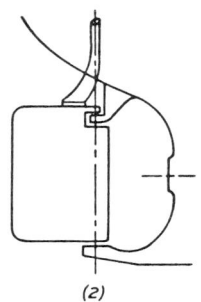 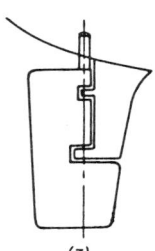

FIG. 16–4—*Types of rudder.*

EXAMPLE 16.5.

The centre line rudder on a twin screw ship is as shown in Figure 16–3. Determine for 35 degrees and a ship speed of 19·5 knots the force and torque on the rudder.

Area A. Force = $15·5 \, A V^2 \, \theta$ Newtons
       = $15·5 \times 2·7^2 \times (19·5 \times 0·5144)^2 \times 35$
       = 397,700 Newtons
Centre of pressure abaft axis = $0·35 \times 2·7 = 0·945$ m
Moment on A_1 = $397,700 \times 0·945$
       = 375,800 Nm Aft
Force on A_2 = $15·5 \times 1·6 \times 5·3 \times (19·5 \times 0·5144)^2 \times 35$
       = 462,600 Newtons
CP from axis = $0·31 \times 5·3 - 2·6$
       = 0·96 m forward of axis
Moment on A_2 = $462,600 \times 0·96$
       = 444,100 Nm Forward
Resultant force on rudder = $397,700 + 462,600 = 860,300 \, N$
Resultant twisting moment on rudder = 375,800 A
                                      444,100 F
                                      ─────────
                                      68,300 Nm Forward

For rudder shapes other than rectangular it is convenient to assess the area and the position of the centre of pressure by integration. Ordinates are measured at common intervals h. For each ordinate a point 0·35 or 0·31—as applicable—of the length of the ordinate from the leading edge is determined. Such points are the centres of pressure of infinitesimally narrow fore and aft strips. From these the area and moments can be calculated and thus the position of the centre of pressure. By taking moments about the top edge of the rudder the vertical position of the centre of pressure can be obtained.

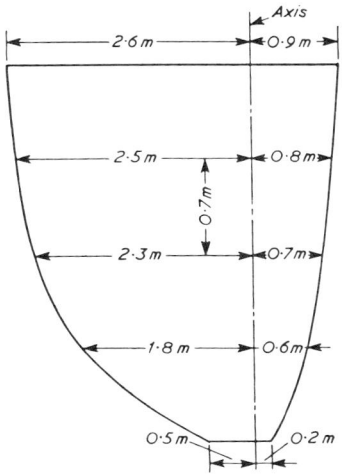

FIG. 16–5—*Example 16.6.*

EXAMPLE 16.6.

A rudder considered as in the open has the outline shown in Figure 16-5. For the ahead condition and an angle of 35 degrees to the middle line determine the position of the centre of pressure relative to the axis of rotation and the top edge of the rudder.

(1)	(2)	(3)	(4)	(5)	(6)	(7)	(8)	(9)	(10)
Ord. Abaft Axis m	Ord. Ford. Axis m	(1) + (2)	SM	fA	C.P. from Ford. (3) × 0·31	C.P. from Axis (6)–(2)	(5)×(7)	Lever	f (5)×(9)
2·6	0·9	3·5	1	3·5	1·09	0·19	0·66	0	—
2·5	0·8	3·3	4	13·2	1·02	0·22	2·90	1	13·2
2·3	0·7	3·0	2	6·0	0·93	0·23	1·38	2	12·0
1·8	0·6	2·4	4	9·6	0·74	0·14	1·34	3	28·8
0·5	0·2	0·7	1	0·7	0·22	0·02	0·01	4	2·8
				33·0			6·29		56·8

Area of rudder $= 33·0 \times \frac{1}{3} \times 0·7$
$= 7·7 \text{ m}^2$

Centre of pressure from axis $= \dfrac{6·29}{33·0} = 0·19$ m abaft axis

Centre of pressure below top $= \dfrac{56·8}{33·0} \times 0·7 = 1·2$ m

Torque

The product of the maximum normal force on the rudder and the distance of the centre of pressure from the axis of rotation gives the net torque which has to be supplied by the steering gear. It should be noted that when the ship is going astern the trailing edge of the rudder becomes the leading edge and the position of the centre of pressure must be measured from this edge.

Diameter of Rudder Stock

Calculations to determine the diameter of the rudder stock need only be made for the maximum helm angle and as stated previously this is usually taken as 35 degrees. There is no advantage to be derived at greater rudder angles. Rudder angles are measured relative to the middle-line plane.

The rudder stock has to transmit the torque and in certain cases such as a spade type rudder the stock may also have a considerable bending moment acting on it.

The normal unbalanced rudder supported on a number of pintles is subjected to considerable twisting moment but with little in the way of bending. With the balanced rudder the number of pintles is reduced to two at the most and hence the bending moment is important. With spade rudders there is no bottom support and the effect of bending could be greater than that of twisting.

The rudder stock diameter also depends upon the material used and the factor of safety adopted. For cast steel the allowable stress is generally accepted as $77 \cdot 2 \times 10^6$ N/m^2.

If the bending moment is *small* then the rudder stock diameter may be determined from the torsion equation.

$$\frac{T}{J} = \frac{f}{d/2}$$

where T = torque in Newton metres
J = polar moment of inertia in m^4 = $\pi d^4/32$
f = allowable stress in N/m^2
d = diameter of stock in metres

so that

$$T = f \frac{\pi d^4}{32} \times \frac{2}{d} = \frac{\pi}{16} f d^3$$

$$\therefore d^3 = \frac{16T}{\pi f}$$

EXAMPLE 16.7.

A single screw ship with a speed of 15 knots has a rectangular rudder 4·6 m deep and 3·05 m wide. The rudder is hung on its forward edge. Determine the diameter of the rudder stock assuming a maximum stress of $77 \cdot 22 \times 10^6$ N/m^2 and a rudder angle of 35 degrees for the ahead condition.

From $Q = 18 \cdot 0 \, AV^2 \, \theta$
$\qquad = 18 \cdot 0 \times 4 \cdot 6 \times 3 \cdot 05 \times (15 \times 0 \cdot 5144)^2 \times 35$
$\qquad = 520,000$ Newtons

Centre of pressure = $0 \cdot 35 \times 3 \cdot 05$
$\qquad\qquad\qquad\qquad = 1 \cdot 07$ m abaft leading edge

$\qquad T = 520,000 \times 1 \cdot 07$ Nm

From $d^3 = \dfrac{16\,T}{\pi f}$

$\qquad d = \sqrt[3]{\dfrac{16 \times 520,000 \times 1 \cdot 07}{\pi \times 77 \cdot 22 \times 10^6}}$

$\qquad\quad = 0 \cdot 332$ m

Bending Moment on Rudder Head

In the unbalanced rudder type (b) as shown in Figure 16–2 in

274 SHIPS AND NAVAL ARCHITECTURE

which the rudder is carried on a number of pintles the bending moment can be ignored.

In the balanced rudder spade type (a) which is supported only at the rudder head there is considerable bending moment which increases the stress produced by the torque.

If the normal force is Q and the distance of the centre of pressure is Y from the support then the bending moment on the rudder head will be QY and this must be associated with the twisting moment as shown later. See Example 16.9.

The balanced rudder type (b) can in general be treated as two rectangles having loading Q_1 and Q_2 and depths h_1 and h_2 as indicated in Figure 16-2. In this type with rigid constraint at rudder head the bending moment at the rudder head is given by:

$$\frac{Q_2 h_2}{4} - \frac{Q_1 h_1}{8}$$ See Example 16.10.

Combined Bending and Twisting

The effect of both bending and twisting at the rudder stock can be determined by using an equivalent twisting moment given by the expression

$$M + \sqrt{M^2 + T^2}$$

where M is the bending moment and T is the twisting moment.

If this is equated to $\pi/16\, f\, d^3$ the diameter of the rudder head to take account of both effects can be determined.

EXAMPLE 16.8.

The centre of pressure of a rudder is 0·21 m abaft the axis of rotation and 1·22 m below the bearing. The normal force on the rudder is 600 KN. Determine the diameter of the rudder stock if the maximum stress allowed is $77·22 \times 10^6\ N/m^2$.

Bending moment $(M) = 600 \times 1,000 \times 1·22 = 732,000\ Nm$

Twisting moment $(T) = 600 \times 1,000 \times 0·21 = 126,000\ Nm$

Equivalent twisting moment $= M + \sqrt{M^2 + T^2}$

$$= 732,000 + \sqrt{732,000^2 + 126,000^2}$$

$$= 1,474,900\ Nm$$

thus $1,474,900 = \pi/16 \times 77·22 \times 10^6 \times d^3$

$$\therefore d = \sqrt[3]{\frac{1,474,900 \times 16}{\pi \times 77·22 \times 10^6}}$$

$$= 0·46\ m$$

EXAMPLE 16.9.

A balanced rudder type (a) as in Figure 16–2 has an area of 7·5 m²

RUDDERS: OSCILLATIONS

with the centre of pressure 0·18 m from the axis and 1·3 m below the coupling. The ship is twin screw with a speed of 15 knots. Maximum angle of helm is 35 degrees. Determine diameter of rudder stock assuming $f = 77 \cdot 22 \times 10^6 \, N/m^2$

$$V = 15 \times 0 \cdot 5144 = 7 \cdot 72 \, m/sec$$

Normal force on rudder $= Q = 15 \cdot 5 \, A \, V^2 \, \theta$
$= 15 \cdot 5 \times 7 \cdot 5 \times (7 \cdot 72)^2 \times 35$
$= 242,500 \, N$

Twisting moment $(T) = 242,500 \times 0 \cdot 18 = 43,650$ Nm
Bending moment $(M) = 242,500 \times 1 \cdot 3 = 315,200$ Nm

Equivalent twisting moment $= M + \sqrt{M^2 + T^2}$
$= 315,200 + \sqrt{315,200^2 + 43,650^2}$
$= 633,300$ Nm

with $f = 77 \cdot 22 \times 10^6$ N/m² and Equivalent twisting moment $= \pi/16 \, fd^3$

$$d = \sqrt[3]{\frac{633,300 \times 16}{\pi \times 77 \cdot 22 \times 10^6}}$$
$= 0 \cdot 347$ m.

Denny put forward an empirical formula for the diameter of the spade type balanced rudder which when modified for metric units takes the form:

$$D = 0 \cdot 0835 \sqrt[3]{A \times \text{arm} \times V^{1 \cdot 25} \times C}$$

where $D =$ diameter of rudder stock in metres
$A =$ area of rudder in m²
arm $=$ distance of centroid of rudder area from the bearing in metres
$C =$ constant, say 0·2
assumed stress $= 77 \cdot 22 \times 10^6$ N/m²

Applying this to the foregoing example:

$$D = 0 \cdot 0835 \sqrt[3]{7 \cdot 5 \times 1 \cdot 3 \times 15^{1 \cdot 25} \times 0 \cdot 2}$$
$= 0 \cdot 0835 \sqrt[3]{57 \cdot 55}$
$= 0 \cdot 322$ m.

EXAMPLE 16.10.

For the rudder shown in Figure 16–3 determine the bending moment on the rudder assuming rigid constraint on the rudder stock at the rudder top.

$$\text{Bending moment} = \frac{Q_2 h_2}{4} - \frac{Q_1 h_1}{8}$$
$= \frac{462,600 \times 1 \cdot 6}{4} - \frac{397,700 \times 2 \cdot 7}{8}$
$= 185,000 - 134,200$
$= 50,800$ Nm

Bow Rudders

As the neutral point, to which reference has already been made, is generally fairly well forward a bow rudder is relatively much less effective. Moreover due to being well forward a bow rudder is exposed to damage. Only a few bow rudders are fitted to ships in service.

In a number of cases a lateral thrust unit at the bow has been fitted. This in effect is a propeller in a transverse tube.

Special Rudders

The orthodox rudder has a limited effect at low speeds and one way of making available additional manoeuvring capability is to deflect the propeller race.

Kitchen Rudder

The deflexion of the propeller race is achieved by the kitchen rudder which consists essentially of two curved plates shrouding the propeller. For full ahead motion the two plates are for all intents and purposes, parallel with the propeller race causing little interference. When the plates are turned in plan they cause the propeller race to be deflected and thus produce a lateral thrust. When the two plates are turned so as to close in the space behind the propeller they reduce the ahead thrust and finally bring about an astern thrust. The kitchen rudder can thus provide lateral thrust at low ahead speed and also vary the magnitude and direction of the propeller thrust. This rudder is in the main used in small craft and low powered craft.

Voith-Schneider Propeller is a vertical axis propeller and consists of a horizontal disc carrying a number of vertical blades of aerofoil shape. As the horizontal disc is rotated about a vertical axis, a special mechanism feathers the blades in such a way as to provide a thrust in any direction desired. The thrust is made to act fore and aft for propulsion and athwartships for steering. The main advantage is good steering at low speeds.

Flettner Rudder

In this the main rudder is of the balanced type and is free to rotate through 360 degrees. It carries a small controllable flap on the aft side—when going ahead. For a turn to port, the flap is set to starboard; this pushes the main rudder to port. The main rudder thus becomes an ordinary balanced set to port and only requiring additional area to compensate for the activating flap area.

Tutin Rudder

This is a combination of contra-propeller and rudder. The upper and lower portions are curved in opposite directions according to the direction of rotation of the propeller. It is claimed there is a useful contribution to thrust.

In view of the ever-increasing size of ships, good manoeuvring ability has become of very great importance. The orthodox rudder is a relatively effective and simple manoeuvring device which has remained more or less unchanged for centuries. However, since the force it generates is proportional to the square of the velocity of flow across it the rudder is most effective when the propeller race is strong and that is at maximum revolutions. Moreover the maximum angle of helm is generally limited to about 35 degrees as at greater angles the rudder can stall and increase in angle gives no real increase in manoeuvrability.

In effect the rudder is least effective in its operation where the requirements are most demanding. The Ship Division of the National Physical Laboratory has made a study of a number of devices with the object of improving the manoeuvrability of ships throughout the whole speed range.

The two mechanical systems that offered the greatest advantages were the Rotating Cylinder Rudder (RCR) and the Rotating Cylinder Flap Rudder (RCFR). Of the configurations tested the rotating cylinder in the leading edge of the rudder seemed the most practical; normal course-keeping being unimpaired.

Oscillations

The principal ship oscillations are

1) *Rolling* about a fore and aft axis
2) *Pitching* about an athwartship axis
3) *Heaving* in a vertical direction.

The oscillations of ships are a feature which if excessive in magnitude can increase stresses in the structure and bring distress to both passengers and crew. When a ship is in a seaway it is subjected to the types of motion stated above although either rolling or pitching will predominate. Each oscillation has its own natural period; periods are reckoned from port to port or from bow down to bow down and measured in seconds.

Unresisted Rolling in Still Water

The period of oscillation T_r when rolling unresistedly in still water is given by

$$T_r = \frac{2\pi \; K_T}{\sqrt{g \; GM}}$$

$$= \frac{2\pi \; K_T}{3 \cdot 133 \; \sqrt{GM}}$$

$$= \frac{2 \cdot 0 \; K_T}{\sqrt{GM}}$$

where K_T = the polar radius of gyration = $B/3$ approximately and where K_T and GM are in metres.

The period is practically constant for angles up to about 10 degrees. Such rolling is termed *isochronous*. The period varies as K_T; it is increased when weights are distant from the axis of oscillation. The period varies inversely as $\sqrt{GM}$. The addition of a weight to a ship alters both K_T and GM.

Pitching

The expression for the period of pitching of a ship is similar in form to that for rolling if the movement of water as the ship oscillates is neglected.

The period of pitching is given by

$$T_P = \frac{2 \cdot 0 \; K_L}{\sqrt{GM_L}}$$

where K_L = the longitudinal radius of gyration in metres about the transverse axis of pitching—generally through the ship's G; $K_L = L/3$ approximately; GM_L = the longitudinal metacentric height in metres.

$$T_P \doteqdot 0 \cdot 45 \; \sqrt{L} \text{ for fine ships}$$
$$\doteqdot 0 \cdot 82 \; \sqrt{L} \text{ for full ships.}$$

The pitching period is normally about $\frac{1}{2}$ to $\frac{2}{3}$ the rolling period.

Heaving

This is the overall vertical movement of the centre of gravity of a ship in a seaway.

$$T_H = 2 \; \sqrt{\frac{\Delta}{TPM}} = 2 \; \sqrt{\frac{V \times 1 \cdot 025}{A \times 1 \cdot 025}} = 2 \; \sqrt{\frac{V}{A}}$$

where TPM = tonnes per metre; V = volume of displacement in m^3; A = area of waterplane in m^2.

The periods of pitching and heaving do not differ greatly and are about half the rolling period.

RUDDERS: OSCILLATIONS

EXAMPLE 16.11.

A ship has a length of 97·5 m; breadth 10·65 m; draught 3·35 m; transverse metacentric height = 0·76 m; longitudinal metacentric height = 110 m.

$K_T = 3·65$ m; $K_L = 24·4$ m; $C_b = 0·54$; $C_w = 0·70$

Determine the natural periods of: 1) rolling, 2) pitching, and 3) heaving.

1) $T_R = \dfrac{2·0\, K_T}{\sqrt{GM_T}} = \dfrac{2·0 \times 3·65}{\sqrt{0·76}} = 8·4$ secs.

2) $T_P = \dfrac{2·0\, K_L}{\sqrt{GM_L}} = \dfrac{2·0 \times 24·4}{\sqrt{110}} = 4·65$ secs.

3) $T_H = 2\sqrt{V/A}$ $\qquad V = 97·5 \times 10·65 \times 3·35 \times 0·54$

$ = 2\sqrt{\dfrac{3·35 \times 0·54}{0·70}} \qquad A = 97·5 \times 10·65 \times 0·70$

$ = 3·2$ secs.

EXAMPLE 16.12.

A ship of 5,600 tonnes displacement has a transverse metacentric height of 0·91 m, a period of 15 seconds and a horizontal curve of metacentres. Determine the period when 25 tonnes are added at a height of 20 m above the ship's centre of gravity.

Original $K_T = \dfrac{T_R \times \sqrt{GM}}{2·0} = \dfrac{15 \times \sqrt{0·91}}{2·0} = \dfrac{15 \times 0·955}{2·0} = 7·15$ m.

Original $MI = \Delta K^2 = 5,600 \times 7·15^2 = 286,000$

Rise in $G = \dfrac{25 \times 20}{5,625} = 0·0889 = x$

$\therefore GM = 0·91 - 0·089 = 0·821$ m.

New MI = original $MI + 25 \times 20^2 -$ new $\Delta \times x^2$

about new G

$ = 286,000 + 10,000 - 5,625 \times 0·0889^2$

$ = 295,950$

$\therefore$ new $K = \sqrt{\dfrac{295,950}{5,625}} = 7·25$ m.

new period $T_R = \dfrac{2·0 \times 7·25}{\sqrt{0·821}} = 16$ seconds

EXAMPLE 16.13.

A ship of displacement 2,540 tonnes and transverse metacentric height of 0·765 m has a period of roll of 8·5 seconds. Estimate the probable period for a similar ship of 3,050 tonnes displacement and metacentric height of 0·825 m.

Basis ship: $T_R = \dfrac{2·0\, K_T}{\sqrt{GM}}\qquad$ For similar ship's linear dimensions $\propto \Delta^{1/3}$

$$8 \cdot 5 = \frac{2 \cdot 0 \times K_T}{\sqrt{0 \cdot 765}} \qquad K_T \text{ for design} = 3 \cdot 718 \left(\frac{3{,}050}{2{,}540}\right)^{1/3}$$

$$= 3 \cdot 718 \times 1 \cdot 063$$

$$\therefore K_T = \frac{8 \cdot 5 \times 0 \cdot 8748}{2 \cdot 0} \qquad \qquad = 3 \cdot 952 \text{ m.}$$

$$= 3 \cdot 718 \text{ m.}$$

$$T_R \text{ for design} = \frac{2 \cdot 0 \times 3 \cdot 952}{\sqrt{0 \cdot 825}}$$

$$= 8 \cdot 70 \text{ secs.}$$

EXAMPLE 16.14.

A ship of 13,700 tonnes displacement has a metacentric height of 2·0 m and a period of 11 secs. 300 tonnes of oil fuel in the double bottom with a centroid 4·6 m below the ship's centre of gravity are distributed equally port and starboard with centroids equi-distant from the middle line and 3·7 m higher than before. Estimate the new period of roll.

Original $K_T = \dfrac{T_R \times \sqrt{GM}}{2 \cdot 0} = \dfrac{11 \times \sqrt{2}}{2 \cdot 0} = 7 \cdot 778$ m.

Original $MI = \Delta\, K^2 = 13{,}700 \times 7 \cdot 778^2$

$\qquad \qquad = 828{,}800$

Rise in $G = \dfrac{300 \times 3 \cdot 7}{13{,}700} = 0 \cdot 0810$ m.

new $GM = 2 \cdot 0 - 0 \cdot 081 = 1 \cdot 919$ m.

New MI = original MI − weight of fuel
  (initial separation of centroid of fuel and G of ship)^2
  + weight of fuel × Y^2 − Δ (movement of G of ship)^2
  = $828{,}800 - 300 \times 4 \cdot 6^2 + 300 \times 0 \cdot 9^2 - 13{,}700 \times 0 \cdot 081^2$
  = $828{,}800 - 6{,}350 + 240 - 90$
  = $822{,}600$

new $K = \sqrt{\dfrac{822{,}600}{13{,}700}} = 7 \cdot 75$ m.

new $T_R = \dfrac{2 \cdot 0 \times 7 \cdot 75}{\sqrt{1 \cdot 919}}$

$\qquad = 11 \cdot 2$ secs.

Y = final separation of centroid of fuel and original G of ship

Ship Stabilization

There is a limit to the extent to which amplitudes of motion of a ship can be reduced by changes in hull shape but considerable reductions in roll amplitudes are possible by other means. From the aspect of comfort, rolling is, in general, the most objectionable of all ship motions.

The resistances to rolling may be stated as follows:
1) The resistance of the air upon the exposed surfaces of the ship
2) The friction of the water upon the immersed surface
3) Eddy-making resistance due to projections on the immersed surface
4) Wave resistance caused by surface disturbance.

Of the various ship oscillations, rolling motion was the first to have a mathematical analysis. Froude in 1865 in a published work on the rolling of ships suggested that bilge keels should be fitted to ships to reduce the rolling motion through damping. Watts experimented with a passive-tank stabilizer about twenty years later. In recent years there has been a substantial increase in the number of ships fitted with some form of stabilizing or roll damping force.

Roll stabilization systems come under two headings (a) passive and (b) active.

Passive Systems

In these systems no separate source of power is required and no special control system.

Bilge Keels

Some such as the bilge keel are external to the main hull and thus create a resistance to ahead motion. The usefulness of bilge keels has often been questioned. Recently experiments were carried out to determine the effect of forward speed on the roll damping coefficient. These indicated that if two models, one with and one without bilge keels, are compared then the absolute difference in damping coefficient is constant for varying speeds and the damping coefficient increases with speed. So that with increasing speed the percentage influence of the bilge keel decreases. In service, ordinary bilge keels produce a roll reduction of about 30 per cent and thus are about half as effective as passive tank systems and a third as effective as activated fins. The size of bilge keels is restricted due to their effect on the ship's speed and the problems that arise in structural connexions, and this in turn makes them only marginally effective as a roll reducing device.

Bilge keels generally extend over the middle half to two-thirds of the ship's length and are fitted at the turn of the bilge.

Other passive systems are:
Fixed fins; passive tank system; passive moving weight system.

Fixed Fins

These are very similar in action to bilge keels with the exception that they extend further from the ship's side and are thus more susceptible to damage. They are generally less effective at low speed.

Passive Tank System

Passive tanks depend upon the oscillation of the liquid between the tanks, port and starboard, being so tuned to the ship's period of roll that the motion of the liquid tends to lag about a quarter of a cycle behind the roll. The object is to arrange the tanks in such a way that the greatest volume of liquid is on the side of the ship which is rising and is, therefore, being raised against gravity. An important requirement of a passive tank system is to ensure that it is effective over a wide range of rolling periods. Such systems normally incorporate some means for adjusting the natural period of oscillation of the liquid to suit the period of roll under different sea conditions.

To obtain satisfactory roll reduction the mass of the liquid in the tanks is generally about three per cent of the displacement. Tank systems bring about some loss of statical stability on account of the free-surface effect of the liquid, but the reduction is generally small. For passive tanks reductions of the order of 60 to 80 per cent have been claimed for angles of roll at, or near, resonance conditions.

Passive Moving Weight System

This is similar in principle to the passive tank system but in general is less effective for a given weight of system. It is effective at low speeds and occupies less space than the tank system.

Active Systems

The active systems are characterized by the fact that control equipment is required to anticipate motion and activate the stabilizer to apply the necessary roll-reducing moments.

As with passive systems the active systems may be internal or external to the main hull. The main active systems are:
Active fins; active tank system; active moving weight; gyroscope.

Activated Fin

Of the active systems the activated fin is the most highly developed and is probably the most widely used system at the present time. The first successful installation was made by Motora in Japan in the 1920s at the same time as a similar system was being developed in the United Kingdom.

This system employs fins projecting from the ship on each side, near the bilge. The fin is connected to a driving mechanism which rotates the fin to various angles. With the ship rolling to port, the port fin is inclined downwards from the leading edge and the starboard fin upwards. Each fin thus contributes a stabilizing moment. On a roll to starboard the slope of the fin is reversed. There are two variations of the activated fin stabilizer: non-retractable and retractable. With the latter the fin when not in use is withdrawn into the hull. The fins are normally designed to give a roll reduction of about 90 per cent. Fig. 16-6 shows a Sperry Gyrofin with the fin rigged out.

FIG. 16-6—*Activated fin stabilizer.*

Active Tank Systems

There are various systems available but in each the essential features are two tanks, one on each side of the ship, in which the water level can be controlled by a sensing system. The active tank system, which is still in the development stage, has the disadvantage that in addition to requiring sensitive control mechanism it has intermittent power requirements with a high peak demand.

Active Weights

This can take several forms but in principle consists of a mass (W) attached to a rotating arm of radius (R). When the arm is at an angle θ to the middle line of the ship and on the high side the

$$\text{Righting moment} = W R \sin \theta$$

Such a system is effective at low speeds but the mass is high and the power requirement large. There could be effective reduction in statical stability if the system stalled with the mass all on one side.

Gyroscopes

Roll stabilization by means of gyroscopes depends upon the high speed rotation of large masses to produce stabilizing couples.

There is a roll reduction of about 50 per cent. It is effective at low speeds but the system is heavy, about 2 per cent of the displacement, has a large power requirement and occupies considerable space in the hull.

Roll stabilization by activated fin, passive tank, and active tank systems—the latter now advanced—are well established and widely adopted.

The methods used to stabilize against rolling can be used to stabilize against pitching but broadly the forces or powers involved are too large to justify their use.

CHAPTER 17

Resistance and Powering

The power required to propel a ship through water depends upon
a) the resistance offered by the water and air
b) the efficiency of the propulsion unit adopted
c) the interaction between them

The total resistance of a ship moving on a calm water surface has a number of components, in wave-making resistance; frictional resistance; form drag; eddy making resistance; air resistance; appendage resistance.

Wave Making Resistance

A body moving on an undisturbed water surface produces a wave system. There are three types of waves generally formed when a ship moves through still water, namely bow and stern divergent and transverse waves as shown in Figure 17–1. The wave system arises

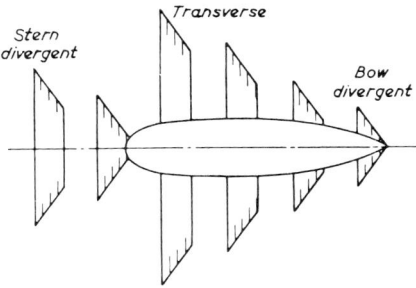

FIG. 17–1—*Types of waves generated by ship motion through water.*

from the pressure field around the ship and the energy possessed by it is derived from the ship. The transfer of energy shows itself as a force opposing forward motion. This force is the wave-making resistance.

Frictional Resistance

When a body moves through a fluid which is otherwise at rest, a thin layer of fluid adheres to the surface of the body and has no velocity relative to the body. At some distance from the body the fluid remains at rest. The variation of velocity of the fluid is rapid close to the body but reduces with increasing distance from the body. This region where there is a rapid change in velocity is termed the boundary layer. Figure 17-2. It is customary to define the thickness which in-

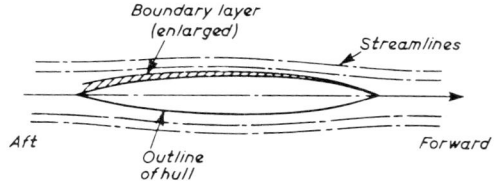

FIG. 17-2—*Boundary layer.*

creases from forward to aft as the distance from the surface of the body at which the velocity of the fluid is one per cent of the body velocity. The body experiences a resistance which is termed the frictional resistance.

Form Resistance. The water particles moving past the hull in their streamlines cannot always exactly follow the ship's form precisely and break away. The pressure acting on the stern is reduced so that there arises a resultant force opposing forward motion. This force is the form drag or resistance.

Eddy-making Resistance. When the streamline flow breaks down, a disturbed volume of water is formed in which water particles revolve in eddies. The energy of this motion is wasted and can be treated as an increase in resistance.

Air Resistance. Air is a fluid and as such will resist the passage of the exposed portions of the ship through it. The resistance has both frictional and eddy-making components. At the full speed of the ship in conditions of no wind the air resistance is about 2 to 4 per cent of the total water resistance. In severe weather the air resistance can contribute appreciably to slowing down the ship.

Appendage Resistance. Typical appendages are rudders, shaft brackets, bossing, stabilizers and bilge keels. Appendage resistances are generally small, in the order of 10 per cent of that of the hull.

In the assessment of ship resistance it is customary to group wave-making resistance, form resistance, eddy resistance and air resistance into one force termed *Residuary Resistance.*

So that the total resistance is given by

$$R_t = R_r + R_f$$

where R_r = residuary resistance and R_f = *frictional resistance.*

Experiments have shown that the frictional resistance between the hull of a ship and the water depends on the area of the surface, the speed of the ship and the degree of roughness of the hull. Experiments also indicate that such resistance is not directly proportional to the length, but that the longer the surface the less is the frictional resistance per unit of length.

Many notable men such as Bouguer, Tiedemann, Newton, Chapman, Euler and Beaufoy made attempts to use models to show how they could be used to predict full-scale behaviour. It was, however, not until the time of William Froude that full-scale prediction became a workable proposition.

It was Froude who put forward the idea of dividing the total resistance into the residuary resistance and the frictional resistance of the equivalent flat plate. By observing the wave patterns created by geometrically similar forms at different speeds, Froude found that the patterns appeared to be geometrically identical when the models were at speeds proportional to the square root of their lengths. This speed is known as the corresponding speed. This with his further work led to *Froude's Law of Comparison* which can be stated thus:

If two geometrically similar forms—two ships or a ship and its model—are run at speeds proportional to the square root of their lengths—corresponding speeds—then their residuary resistances per unit of displacement will be the same.

This is an important law which makes it possible to estimate the residuary resistance of a ship from that of her model or from that of a ship of different size but of the same form.

The Froude tank constructed in 1871 was for all intents and purposes a large trench cut along a hillside at Chelston Cross near Torquay. The waterway was about 76 m in length with a breadth of 2 m at the bottom and 8 m at the top. The depth was about 3 m. That tank, the world's first model tank has been followed by many throughout the world, all contributing in greater or less degree to the elucidation of the problems which have developed over the years.

Modern ship tanks for measuring model resistance are basically the same as the first tank constructed for Froude. They are, for present day requirements, a long tank of more or less rectangular cross-section, spanned by a carriage which tows the model along the tank. Figure 17–3. In a typical run the carriage is accelerated to the required speed, resistance records are made by a dynamometer and records of trim of the model are made by recording pens during a period of constant speed, and then the carriage is decelerated.

The models vary in length from about 4 to 9 metres. In the United Kingdom they are generally of wax as wax models are easily shaped and what is also important after tests with a wax model have been completed the model can be melted down and the material used again. Prior to a test the model is ballasted so that it floats at the desired draught and trim.

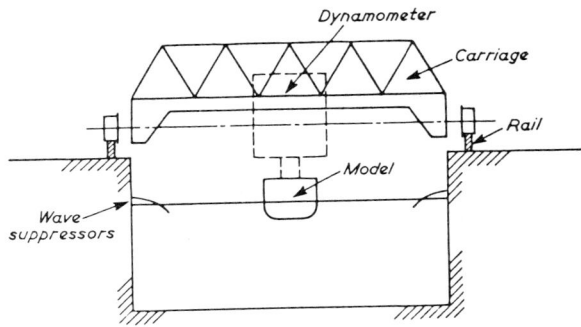

FIG. 17-3—*Section of experimental tank.*

Model Tests

To determine the total resistance (R_t) of a ship from the results obtained in a model test, the procedure is as follows:
1) Measure the total resistance (r_t) of a geometrically similar model at its corresponding speed.
2) Estimate the frictional resistance r_f of the model using the expression given later. The value determined should be corrected for density.
3) Determine the value of the residual resistance r_r of the model from the expression $r_r = r_t - r_f$.
4) By Froude's Law of Comparison determine the value of R_r for the ship in sea water. $R_r = r_r (L/l)^3$.
5) Estimate the value of R_f for the ship by the expression given later.
6) Determine the value of R_t for the ship using $R_t = R_f + R_r$.

The total resistance so determined represents the resistance of the smooth or naked hull and does not include appendages such as bossing etc.

Frictional Resistance calculation

Froude in his investigations into frictional resistance concluded that it depends upon:
a) the area of the surface
b) the type of surface
c) the length of the surface
d) the density of the water
e) the nth power of the speed.

Froude found that frictional resistance could be expressed by the formula

$$R_f = f S V^{1 \cdot 825}$$

where R_f = frictional resistance in Newtons
 f = a coefficient depending on length of surface
 S = wetted surface area of ship in m²
 V = speed of ship in m/s.

The values of f in sea water are given in Table 17–1.

Table 17.1.
f values. R. E. Froude's skin friction constants.

Length (m)	f	Length (m)	f	Length (m)	f
2·0	1·966	11	1·589	40	1·464
2·5	1·913	12	1·577	45	1·459
3·0	1·867	13	1·566	50	1·454
3·5	1·826	14	1·556	60	1·447
4·0	1·791	15	1·547	70	1·441
4·5	1·761	16	1·539	80	1·437
5·0	1·736	17	1·532	90	1·432
5·5	1·715	18	1.526	100	1·428
6·0	1·696	19	1·520	120	1·421
6·5	1·681	20	1·515	140	1·415
7·0	1·667	22	1·506	160	1·410
7·5	1·654	24	1·499	180	1·404
8·0	1·643	26	1·492	200	1·399
8·5	1·632	28	1·487	250	1·389
9·0	1·622	30	1·482	300	1·386
9·5	1·613	35	1·472	350	1·376
10·0	1·604				

For $R_f = f S V^{1.825}$ where R_f = frictional resistance in Newtons
 S = wetted surface in m²
 V = speed in m/s
 L = ship length in m

Wetted Surface

The wetted surface area of a ship may be estimated by the following expression

$$S = C \sqrt{\Delta L}$$

where S = wetted surface area in m²
 Δ = displacement of ship in tonnes
 L = length of ship in m
 C = a coefficient, generally about 2·58.

Froude's Law of Comparison

This, as stated above, is an important law. The law states that the residuary resistance of a ship and her model vary as the ratio of the displacements or (lengths)^3 when the ship and the model are run at corresponding speeds.

The corresponding speed is that speed which makes:

$$\frac{V}{\sqrt{L}} = \frac{v}{\sqrt{l}}$$

where V and L are for the ship and v and l are for the model.

also $\quad \dfrac{V}{\sqrt{L}} = \dfrac{v}{\sqrt{l}} \quad$ or $\quad \dfrac{V}{v} = \sqrt{\dfrac{L}{l}}$

Now
$$\frac{W}{w} = \left[\frac{L}{l}\right]^3 \quad \therefore \quad \frac{L}{l} = \left[\frac{W}{w}\right]^{1/3}$$

W and w are displacements in tonnes.

Hence
$$\frac{V}{v} = \sqrt{\left(\frac{W}{w}\right)^{1/3}} = \left(\frac{W}{w}\right)^{1/6}$$

EXAMPLE 17.1

(a) A ship of 16,000 tonnes displacement has a speed of 20 knots. Determine the corresponding speed for a similar ship of displacement 12,800 tonnes.

$$\frac{V}{v} = \left(\frac{W}{w}\right)^{1/6}$$

$$\therefore \quad v = V\left(\frac{w}{W}\right)^{1/6} = 20\left(\frac{12800}{16000}\right)^{1/6}$$

$$= 19 \cdot 27 \text{ knots}$$

(b) A ship of length 137 m has a speed of 16 knots. Determine the corresponding speed for a model of length 5 m.

$$\frac{V}{\sqrt{L}} = \frac{v}{\sqrt{l}}$$

$$\therefore \quad v = V\sqrt{\frac{l}{L}}$$

$$= 16 \times \frac{2 \cdot 236}{11 \cdot 70}$$

$$= 3 \cdot 21 \text{ knots}.$$

In symbols
$$\frac{R_r}{r_r} = \frac{W}{w} = \frac{L^3}{l^3} \quad \text{if} \quad \frac{V}{\sqrt{L}} = \frac{v}{\sqrt{l}}$$

where W and w are displacement in tonnes; R_r and W are for ship; and r_r and w are for model.

Effect of Density

If the model is run, as is highly probable, in fresh water, a correction must be made for the fact that the ship will be in sea water.

Assuming that resistance varies as the density then if x is the resistance in fresh water (1 tonne/m³) the residuary resistance in sea water (1·025 tonne/m³) would be $x \times 1\cdot025$.

Powering of Ships

Once the total resistance for a ship has been assessed it is then necessary to express this in terms of power. In metric units 1 horse power is equal to 745·7 watts. The power of a ship is expressed in kW. Power is the rate of doing work.

In propelling a ship through the water the effective work done is in overcoming the resistance of the ship brought about by its speed of advance. The resistance in this case is conventionally taken to be that of the naked hull, that is, without any appendages.

So that the effective power (P_E) of a ship is the product of the resistance of the naked hull and the speed of the hull. Thus

$$P_E = R_t \times V \times 0\cdot5144$$
$$= kN \times \text{metre/sec}$$
$$= kW$$

where R_t is in kN and V is in knots. 1 knot $= 0\cdot5144$ m/sec.

The various ship powers are denoted by the standard nomenclature as follows:

$P_B =$ Brake
$P_S =$ Shaft
$P_D =$ Delivered
$P_T =$ Thrust

These are dealt with in Chapter 18.

EXAMPLE 17.2

The total resistance of a ship is 945 kN at a speed of 17 knots. Determine the effective power.

$$P_E = 945 \times 17 \times 0\cdot5144$$
$$= 8250 \text{ kW}$$

Note 1 watt $=$ 1 joule/sec. $=$ 1 Nm/sec.

EXAMPLE 17.3.

A ship of 5100 tonnes displacement has a speed of 20 knots, a length of 120 m and is to be tank tested using a model of length 3 m. Determine the speed at which the model should be tested and the ratio of the ship to model P_E at this speed.

$\Delta \propto L^3 \quad \therefore \quad \Delta \text{ model} = 5100 \, [3/120]^3 = 0\cdot079 \text{ tonne.}$

$$\frac{V}{\sqrt{L}} = \frac{v}{\sqrt{l}}$$

$$\frac{20}{\sqrt{120}} = \frac{v}{\sqrt{3}}$$

$$\therefore \quad v = 3{\cdot}15 \text{ knots}$$

Now

P_E at corresponding speed for similar ships $\propto [\Delta]^{7/6}$

Ratio of P_E ship to model $= \left(\dfrac{5100}{0{\cdot}079}\right)^{7/6} = 64550^{7/6}$
$= 4{\cdot}08 \times 10^5$

EXAMPLE 17.4.

A model 5·2 m in length has a total resistance of 67 newtons when towed at 3·5 knots. The wetted surface of the model is 4·25 m². Determine the P_E for the ship of similar form and length 131 m at the corresponding speed.

$$f \text{ model} = 1{\cdot}728 \text{ sw}$$
$$f \text{ ship} = 1{\cdot}418 \text{ sw}$$
$$3{\cdot}5\text{K} = 3{\cdot}5 \times 0{\cdot}5144$$
$$= 1{\cdot}8 \text{ m/s}$$

$r_f = fsv^{1{\cdot}825} = 1{\cdot}728 \times 4{\cdot}25 \times 1{\cdot}8^{1{\cdot}825} \times \dfrac{1}{1{\cdot}025} = 21 \text{ N}$

$r_r = 67 - 21 = 46 \text{ N}$

For ship

Corresponding speed $= 3{\cdot}5\sqrt{\dfrac{131}{5{\cdot}2}} = 17{\cdot}6 \text{ knots} = 17{\cdot}6 \times 0{\cdot}5144$
$= 9{\cdot}05 \text{ m/s}$

$R_f = 1{\cdot}418 \times 4{\cdot}25 \times \left(\dfrac{131}{5{\cdot}2}\right)^2 \times 9{\cdot}05^{\,1{\cdot}825}$

$= 213500 \text{ N}$

$R_r = 46 \times \left(\dfrac{131}{5{\cdot}2}\right)^3 \times 1{\cdot}025 = 754000 \text{ N}$

$R_t = 213500 + 754000 \qquad = 967600 \text{ N} = 967{\cdot}6 \text{ kN}$

$P_{En} = 967{\cdot}6 \times 17{\cdot}6 \times 0{\cdot}5144 \qquad = 8750 \text{ kW}$

EXAMPLE 17.5.

A ship of length 137 m has a wetted surface of 3,700 m². Determine the frictional residual of this ship at 14 knots. Assess, also, the residual resistance of the ship at that speed if the residuary resistance of its model of 4·6 m length is 3·5 N in sea water.

Finally determine the P_E for the ship at 14 knots. f ship $= 1{\cdot}416$

$R_f = fSV^{1{\cdot}825} = 1{\cdot}416 \times 3700 \times 7{\cdot}2^{1{\cdot}825} \qquad 14\text{K} = 14 \times 0{\cdot}5144$
$= 192500 \text{ N} \qquad\qquad\qquad\qquad\qquad\quad = 7{\cdot}2 \text{ m/s}$

RESISTANCE AND POWERING 293

$$R_r = 3.5 \left(\frac{137}{4.6}\right)^3 = 92430 \text{ N}$$

∴ $R_t = 192500 + 92430 = 284930 \text{ N} = 284.93 \text{ kN}$
   $P_E = 284.93 \times 14 \times 0.5144$
   $= 2050 \text{ kW}$

EXAMPLE 17.6

The power calculation for a ship of length 119 m and displacement 9,700 tonnes gives a naked P_E value of 1,750 kW for a speed of 12 knots. The power is to be checked by running a model 4·55 m long in an experimental tank. Determine the pull applied to the model at the corresponding speed.

   f model $= 1.760$ FW f ship $= 1.422$ SW

For Ship

Wetted surface $= 2.58 \sqrt{\Delta L} = 2.58 \sqrt{9700 \times 119} = 2770 \text{ m}^2$

$R_f = fSV^{1.825}$
$= 1.422 \times 2770 \times 6.17^{1.825}$ 12 K $= 6.17$ m/s
$= 109300 \text{ N} = 109.3 \text{ kN}$

$P_E = R_t \times V \times 0.5144$

$R_t = \dfrac{P_E}{V \times 0.5144} = \dfrac{1750}{12 \times 0.5144} = 283.4 \text{ kN}$

$R_t = R_r + R_f$
$R_r = 283.4 - 109.3 = 174.1 \text{ kN}$

For Model

Corresponding speed $= \dfrac{V\sqrt{l}}{\sqrt{L}} = \dfrac{12 \times \sqrt{4.55}}{\sqrt{119}} = 2.35$ knots $= 1.21$ m/s

Wetted surface $= 2770 \left(\dfrac{4.55}{119}\right)^2 = 4.05 \text{ m}^2$

$r_f = 1.760 \times 4.05 \times 1.21^{1.825}$
$= 9.82 \text{ N}$

$r_r = 174.1 \times \left(\dfrac{4.55}{119}\right)^3 \times \dfrac{1000}{1025} \times 1000 = 9.55 \text{ N}$

∴ $r_t = r_f + r_r = 9.82 + 9.55$
$= 19.37$ N.

EXAMPLE 17.7.

A ship has a length of 122 m, breadth 19·8 m, draught 7·33 m and a displacement of 8,700 tonnes. A model of the ship is run in a tank with the following results:

V	16	17	18	19	20	Knots
P_E	2420	3010	3740	4620	5710	kW

Estimate (1) the P_E for a ship of similar form of 16.250 tonnes displacement at a speed of 19·5 knots; and (2) the dimensions of the new design

$$\text{Corresponding speed of basis} = 19\cdot5 \left(\frac{8700}{16250}\right)^{1/6} = 17\cdot57 \text{ K.}$$

P_E of basis at 17·57 knots from curve = 3420 kW.

$$P_E \text{ of design at 19·5 knots} = 3420 \times \frac{16250}{8700} \times \frac{19\cdot5}{17\cdot57}$$
$$= 7100 \text{ kW.}$$

As $L \propto \Delta^{1/3}$ For new design $L = 122 \times \left(\frac{16250}{8700}\right)^{1/3}$

$$= 122 \times 1\cdot231 = 150\cdot2 \text{ m.}$$
$$B = 19\cdot8 \times 1\cdot231 = 24\cdot4 \text{ m.}$$
$$H = 7\cdot33 \times 1\cdot231 = 9\cdot0 \text{ m.}$$

The Reynolds Number

When a flat smooth surface such as a plate moves through a viscous fluid, the fluid near it is carried along with it. Some distance out from the surface the fluid is not affected by the movement of the plate and remains at rest. The fluid set in motion between the surface of the plate and the undisturbed outer liquid is called the boundary layer or frictional wake. The movement of water inside the wake may follow two patterns and it is assumed in both cases that the actual water next to the surface has no motion relative to the surface but is carried along with it. Firstly, if between this and the outer edge of the wake, where the fluid is at rest, the water moves in a series of layers, without mixing, the flow is said to be *laminar*. Secondly, if the flow is such that eddying occurs in the wake, with resulting mixing of the layers, the flow is said to be *turbulent*.

The flow that will be set up in given circumstances depends upon the relative importance of inertia and viscous forces. The former favours turbulent flow, the latter laminar. It can be shown that the ratio of these two forces is represented by the parameter

$$\frac{VL}{\gamma}$$

where V = speed of the surface relative to still water, L = length of surface in direction of motion and γ = coefficient of kinematic viscosity of the fluid. This parameter is called the Reynolds number (Rn) of the motion; that is

$$Rn = \frac{VL}{\gamma}$$

Professor Osborne Reynolds was the first to demonstrate the fundamental differences between laminar and turbulent flow.

Small values of Rn will be associated with laminar flow and large values with turbulent flow.

For S.I. metric units the Reynolds number is given by

$$Rn = \frac{VL}{\gamma} = 0.8415 \times V \times L \times 10^6 \text{ for sea water at 15°C only}$$

where V is in m/sec; and L is in metres.
or $Rn = 0.4329 \times V \times L \times 10^6$
where V is in knots; and L is in metres.

When the kinematic viscosity (γ) values for fresh and sea water are associated with temperatures in deg. Celsius and the values are in centistokes

$$cst = 10^{-6} \text{ m}^2/\text{sec and } \gamma = 1.140 \ cst \text{ for FW}$$

$$\gamma = 1.190 \ cst \text{ for SW}$$

then

$$Rn = \frac{VL}{\gamma} = \frac{V \text{ in m/sec} \times L \text{ in m} \times 10^6}{1.190 \text{ m}^2/\text{sec}} \text{ for SW at 15°C.}$$

so that at 15°C the Reynolds number for a ship of length 122 m and a speed of 15 knots is given by

$$Rn = \frac{15 \times 0.5144 \times 122 \times 10^6}{1.190} = 7.9 \times 10^8$$

Rn is dimensionless and varies with temperature because of changes in γ.

The essential requirement for dynamical similarity between resistance forces due entirely to inertia is that the function

$$\varphi \left(\frac{gL}{V^2} \right)$$

remains constant. The reciprocal function $V/\sqrt{gL}$ is known as the Froude number Fn and for S.I. metric units

$$Fn = 0.3193 \ \frac{V}{\sqrt{L}} \quad \text{where } V \text{ is in m/sec; } L \text{ is in metres.}$$

$$= 0.1643 \ \frac{V}{\sqrt{L}} \quad \text{where } V \text{ is in knots and } L \text{ is in metres.}$$

Lord Rayleigh demonstrated a connexion between Fn and Rn and subsequently a number of investigators assumed that Rn should form the basis of friction calculations in lieu of the Froude empirical formulae. There is a relationship between the coefficient of frictional resistance C_F and Rn and at the International Tank Towing Conference (ITTC) in Madrid in 1957 an agreed friction line was adopted which gives the following:

$$C_F = \frac{0.075}{[\log Rn - 2]^2} \qquad \text{also } C_F = \frac{R_F}{\frac{1}{2}\rho S V^2}$$

and

$$C_T = \frac{R_T}{\frac{1}{2}\rho S V^2}$$

where

ρ = density of the fluid $\begin{cases} 1000 \text{ kg/m}^3 \text{ for FW} \\ 1025 \text{ kg/m}^3 \text{ for SW} \end{cases}$

S = wetted surface in m^2
V = velocity in m/sec.

Table 17.2 gives values for this ITTC formula:

Table 17.2

Values of $C_F \times 10^3$ in ITTC Formula $C_F = \dfrac{0.075}{[\log Rn - 2]^2}$

Reynolds No.	C_F for $Rn \times 10^6$	C_F for $Rn \times 10^7$	C_F for $Rn \times 10^8$	C_F for $Rn \times 10^9$
1·00	4·688	3·000	2·083	1·531
1·25	4·468	2·887	2·018	1·489
1·50	4·301	2·799	1·966	1·456
1·75	4·166	2·728	1·924	1·430
2·00	4·054	2·669	1·889	1·407
2·25	3·960	2·618	1·859	1·387
2·50	3·878	2·574	1·832	1·370
2·75	3·806	2·535	1·809	1·355
3·00	3·742	2·500	1·788	1·342
3·50	3·632	2·440	1·751	1·318
4·00	3·541	2·390	1·721	1·298
4·50	3·464	2·347	1·694	1·280
5·00	3·397	2·309	1·671	1·265
5·50	3·338	2·276	1·651	1·252
6·00	3·285	2·246	1·633	1·240
6·50	3·238	2·220	1·616	1·229
7·00	3·195	2·195	1·601	1·219
7·50	3·156	2·173	1·587	1·209
8·00	3·120	2·152	1·574	1·201
8·50	3·087	2·133	1·562	1·193
9·00	3·056	2·115	1·551	1·185
9·50	3·027	2·099	1·540	1·178

RESISTANCE AND POWERING

EXAMPLE 17.8

A ship has a length of 134 m and a speed of 16 knots. Determine (1) the value of R_F if $\gamma = 1\cdot19 \times 10^6$ and the wetted surface is 3170 m²; and (2) the value of the Froude number Fn.

$$Rn = \frac{VL}{\gamma} = \frac{16 \times 0\cdot5144 \times 134 \times 10^6}{1\cdot19} = 9\cdot25 \times 10^8$$

$$C_F = \frac{0\cdot075}{[\log Rn - 2]^2} = \frac{0\cdot075}{[6\cdot966]^2} = 0\cdot00154$$

$$R_F = C_F \times \tfrac{1}{2}\rho S V^2$$
$$= 0\cdot00154 \times \tfrac{1}{2} \times 1000 \times 3170 \times 8\cdot23^2$$
$$= 165500 \text{ N}$$

$$Fn = 0\cdot3193 \frac{V}{\sqrt{L}} = 0\cdot3193 \times \frac{8\cdot23}{11\cdot57} = 0\cdot227.$$

The following example shows the procedure in the determination of P_E by the ITTC method.

EXAMPLE 17.9.

As Example 17.4 allowing 15 per cent for ship roughness.

Model
Wetted surface = 4·25 m²
$v = 1\cdot8$ m/sec
$\gamma_t = 67$ N
$l = 5\cdot2$ m

Ship
Wetted surface = 2740 m²
$V = 9\cdot05$ m/sec
$L = 131$ m

Kinematic viscosity = $1\cdot140$ *cst* FW
= $1\cdot190$ *cst* SW
Density: FW = 1000 kg/m³
SW = 1025 kg/m³

$$\text{Model } C_T = \frac{R_T}{\tfrac{1}{2}\rho S V^2}$$

$$= \frac{67 \times 2}{1000 \times 4\cdot25 \times 1\cdot8^2}$$

$$= 9\cdot71 \times 10^{-3}$$

$$\text{Model } Rn = \frac{VL}{\gamma} = \frac{1\cdot8 \times 5\cdot2 \times 10^6}{1\cdot140}$$

$$= 8\cdot22 \times 10^6$$

$$\text{Model } C_F = \frac{0\cdot075}{[\log Rn - 2]^2}$$

$$= \frac{0\cdot075}{[6\cdot9150 - 2]^2}$$

$$= \frac{0.075}{4.915^2}$$
$$= 3.10 \times 10^{-3}$$
$$C_R = C_T - C_F$$
$$= [9.71 - 3.10]10^{-3}$$
$$= 6.61 \times 10^{-3}$$

Ship

$$V = 9.05 \text{ m/sec}$$
$$Rn = \frac{9.05 \times 131}{\gamma}$$
$$= \frac{9.05 \times 131 \times 10^6}{1.190}$$
$$= 9.9 \times 10^8$$
$$C_F = \frac{0.075 \times 1.15}{[8.996 - 2]^2} \qquad 15\% \text{ for ship roughness}$$
$$= \frac{0.0863}{6.996^2}$$
$$= 1.77 \times 10^{-3}$$
$$C_T = C_R + C_F \qquad C_R \text{ as model}$$
$$= [6.61 + 1.77]10^{-3}$$
$$= 8.38 \times 10^{-3}$$
$$R_S = C_T \times \tfrac{1}{2}\rho S V^2$$
$$= 8.38 \times 10^{-3} \times \tfrac{1}{2} \times 1025 \times 2740 \times 9.05^2$$
$$= 963000 \text{ N}$$
$$P_E = 963000 \times 9.05 \text{ Nm/sec} = 8750 \text{ kW}.$$

Estimation of Power

In the early stages of a ship design the results of model tests are not generally available and in order to assess the power required for the speed specified it is necessary to make recourse to approximate formulae or methods. Some of these are as follows:

1) *Admiralty Coefficient* $[A_c]$

$$A_c = \frac{\Delta^{2/3} \times V^3}{P}$$

where Δ = displacement in tonnes
V = speed in knots
P = power as P_S or P_B

As a first approximation to the power of a ship the expression

RESISTANCE AND POWERING

given above may be used. The values of the coefficient remain similar for similar ships in their normal speed range. The coefficient should be assessed from basic ship data.

A ship of 15,000 tonnes displacement at 14 knots has a P_B of 3,560 kW then

$$A_c = \frac{15000^{2/3} \times 14^3}{3560} = \frac{608 \times 2744}{3560} \doteqdot 470$$

The expression for A_c is derived as follows:

If resistance $R \propto$ wetted surface, and wetted surface $\propto$ [linear dimensions]^2

then $\qquad S \propto L^2 \propto \Delta^{2/3}$

∴ $\qquad R \propto \Delta^{2/3}$ if V is constant.

If Δ is constant $R \propto V^2$

∴ $\qquad R \propto \Delta^{2/3} V^2$ when V and Δ are variable

also $\qquad P \propto R \times V$

$\qquad P \propto \Delta^{2/3} V^2 \; V \propto \Delta^{2/3} V^3$

Note: $\Delta \propto L^3; \quad L^3 \propto \Delta; \; L \propto \Delta^{1/3}; \quad L^2 \propto \Delta^{2/3}$.

2) *Model Tests.*

The resistance derived from model tests is frequently converted into a power constant known as "circular C" and represented by ⓒ.

This is plotted as

$$ⓒ = 579 \cdot 7 \frac{P_E}{\Delta^{2/3} V^3}$$

or $\qquad P_E = \dfrac{\Delta^{2/3} V^3 \; ⓒ}{579 \cdot 7}$

where P_E = effective power in kW; Δ = displacement in tonnes; and V = speed in knots.

Typical ⓒ values are given below taken from tank model experiments of a ship of 33,300 tonnes displacement.

V	15	16	17	18	Knots
ⓒ	0·741	0·749	0·755	0·811	
From which is obtained					
naked P_E	4470	5490	6640	8460	kW

3) *Taylor's Standard Series.*

This series in the past has been extensively used for estimating the resistance of fast ships. In this method the frictional and residuary resistances are treated separately and both are expressed in lb/per ton of displacement. The results of the series have been re-analyzed by Gertler and published by the USA Government Printing Office.

4) *EHP of Single Screw Ships by Moor*
Details are given in TRINA Vol. 102.
The results are presented in the form of charts and tables of © for a ship of standard dimensions. The information is given for a series of speeds and a range of LCB positions and block coefficients.

CHAPTER 18

Propulsion and Propellers

The power developed by an engine is measured in a particular way:

Steam Turbines. This is measured in terms of shaft power (P_s). The measurement is carried out as near to the propeller as possible by means of a torsionmeter. This instrument measures the angle of twist caused in a specified length of the shaft by the applied torque.

$$P_s = \frac{2\pi Q N}{1000}$$

where
P_s = kW
Q = Newton metres
N = Revs per second

Internal Combustion. This is measured at the crank shaft by a dynamometer (P_B).

Other powers are as follows:

Thrust Power (P_T) This is the product of the thrust (T) delivered to the propeller and the speed of advance (V_a) of the propeller.

$$P_T = T \times V_a$$

where
P_T = kW
T = kN
V_a = m/sec.

Indicated Power. This is the power obtained for each cylinder and when stated the mechanical efficiency of the engine should be included.

Delivered Power (P_D). This is the power delivered to the propeller. It is rather less than P_S due to friction in the bearings.

$$P_D = \frac{2\pi Q N}{1000} \quad \text{as in } P_S$$

and as stated in Chapter 17.

Effective Power (P_E). The power required to tow the ship.

$$P_E = R_t \times V \times 0.5144$$

where
R_t = kN
V = knots
P_E = kW

The various powers are related as follows:
P_B — friction losses in gearing and bearings = P_S
P_S — friction losses in shaft bearings abaft torsion meters = P_D
P_D — propeller losses = P_T
P_T — hull losses = P_E

The overall efficiency of propulsion depends upon
1) efficiency of propelling machinery
2) hull efficiency
3) propeller efficiency.

The ratio between the P_S and P_E is a measure of the overall propulsive efficiency achieved and is termed the propulsive coefficient.

$$\text{The hull efficiency} = \frac{P_E}{P_T}$$

$$\text{The propeller efficiency} = \frac{P_T}{P_D}$$

$$\text{Quasi propulsive coefficient} = \text{QPC} = \frac{P_E}{P_D}$$

$$= \frac{P_E}{P_T} \times \frac{P_T}{P_D}$$

= hull efficiency × propeller efficiency.

The above are dealt with in more detail later.

EXAMPLE 18.1.
At 250 m/minute the tow rope pull of a naked hull is 35·5 kN. Determine the P_E of the hull at this speed.

$$P_E = \text{Pull in kN} \times \text{speed in m/sec}$$
$$= 35\cdot 5 \times \frac{250}{60}$$
$$= 147 \text{ kW}$$

EXAMPLE 18.2.
The thrust of a propeller is 37·5 kN and the speed of the water at the propeller is 11·5 knots. Determine the P_T.

$$P_T = T \times V_a$$
$$= 37\cdot 5 \times 11\cdot 5 \times 0\cdot 5144$$
$$= 222 \text{ kW}$$

EXAMPLE 18.3.
The undernoted particulars are related to a model propeller:
Thrust = 185 N; Torque = 10·2 Nm; speed of advance = 6 knots; rev/min = 720. Determine P_T, P_D and efficiency.

$$P_T = \frac{185}{1000} \times 6 \times 0.5144$$
$$= 0.57 \text{ kW}$$
$$P_D = \frac{2\pi QN}{1000}$$
$$= \frac{2\pi \times 10.2 \times 720}{1000 \times 60} = 0.77 \text{ kW}$$

Propeller efficiency $= \dfrac{P_T}{P_D} = \dfrac{0.57}{0.77} = 74$ per cent

EXAMPLE 18.4.

The indicated power of a ship, where the mechanical efficiency of the propelling unit is 85 per cent, is 8950 kW. Determine the P_E assuming that losses due to gearing etc. = 5 per cent; losses in shaft bearings = 1 per cent and that the QPC = 0.70.

$$P_B = 8950 - \frac{15 \times 8950}{100} = 7610 \text{ kW}$$

$$P_S = 7610 - \frac{5 \times 7610}{100} = 7230 \text{ kW}$$

$$P_D = 7230 - \frac{1 \times 7230}{100} = 7160 \text{ kW}$$

$$P_E = 7160 \times 0.70 = 5010 \text{ kW}.$$

It is self-evident that it is necessary to provide a propulsive device to drive the ship through the water. The force required to propel the ship must be obtained from a reaction against the water by causing a stream of water to move in the opposite direction. A number of ways are available to produce this stream of water aft but certainly the most commonly used is the screw propeller.

Wake

When a propeller rotates it sucks water into itself and discharges it in a well defined slipstream. Due to the rotation of the blades the fluid pressure immediately behind the propeller is increased and the effect of this is to increase the velocity of the mass of water in the slipstream. The change of momentum in this mass provides the propeller thrust. As the ship moves through the water, the friction of the water on the surface of the hull causes a surrounding layer of water to follow in the direction of motion. This belt of water is called the wake and as a result of its effect the speed of the propeller through the wake is generally less than the speed of the ship.

The wake velocity varies in magnitude and direction but is assumed as a matter of convenience, to have a constant forward velocity given by

where V = speed of ship
V_a = speed of advance of the propeller

Wake speed can be expressed as a percentage of either the speed of advance (V_a) of the propeller or the speed (V) of the ship. Froude selected the former whilst Taylor chose the latter as shown below.

Froude

Wake fraction $= w_f = \dfrac{V - V_a}{V_a}$

Hence $V_a = \dfrac{V}{1 + w_f}$

or $V = V_a(1 + w_f)$

Taylor

Wake fraction $= w_t = \dfrac{V - V_a}{V}$

Hence $V_a = V(1 - w_t)$

or $V = \dfrac{V_a}{1 - w_t}$

The relationship between w_f and w_t can be found from the equations.

$$w_f = \frac{w_t}{1 - w_t} \quad \text{and} \quad w_t = \frac{w_f}{1 + w_f}$$

The Froude and Taylor wake fractions are merely different ways of expressing the same phenomenon.

Expressions frequently used in the estimating stages of a design to assess the wake speed (Vw) are

Single Screw
$Vw = (0 \cdot 5 C_b - 0 \cdot 05)V$

Twin Screw
$Vw = (0 \cdot 5 C_b - 0 \cdot 2)V$

EXAMPLE 18.5.

A single screw ship having a block coefficient of 0·70 was found to have a speed on trial of 16·06 knots. The Froude wake fraction was estimated at 0·43. Estimate the speed of the wake.

1) $Vw = V - V_a$

   $= 16 \cdot 06 - 11 \cdot 23$

   $= 4 \cdot 83$

   $V_a = \dfrac{V}{1 + wf}$

   $= \dfrac{16 \cdot 06}{1 \cdot 43}$

   $= 11 \cdot 23$ knots

2) $Vw = (0 \cdot 5\, C_b - 0 \cdot 05)\, V$
   $= (0 \cdot 5 \times 0 \cdot 70 - 0 \cdot 05)\, 16 \cdot 06$
   $= 0 \cdot 30 \times 16 \cdot 06$
   $= 4 \cdot 82$ knots

Thrust Deduction

Due to the reduction in pressure on the after part of the hull

PROPULSION AND PROPELLERS

brought about by the working of the propeller a suction effect is created. This can be treated either as an augment to the resistance of the hull or as a deduction from the thrust of the propellers.

Let R = resistance of ship in Newtons at speed V without propellers
T = thrust of propeller in Newtons

Then $T - R$ = additional resistance caused by action of propeller $(T - R)/R$ is called the augment of resistance and is denoted by a so that augment of resistance equals

$$a = (T - R)/R = T/R - 1 \quad \text{and} \quad 1 + a = T/R.$$

$(T - R)/T$ is called the thrust deduction factor and is denoted by t so that thrust deduction factor equals

$$t = \frac{T - R}{T} = 1 - \frac{R}{T} \quad \text{and} \quad 1 - t = \frac{R}{T}$$

a and t are not independent factors but are connected by the relation

$$(1 + a)(1 - t) = \frac{T}{R} \times \frac{R}{T} = 1$$

Hull Efficiency

The product $(1 + w_f)(1 - t)$ is called the hull efficiency and is an integral part of the make up of the quasi-propulsive coefficient.

This follows from

$$\text{Hull Efficiency} = \frac{P_E}{P_T} = \frac{R \times V}{T \times Va}$$

But $\quad T = R(1 + a) \quad \text{and} \quad Va = \dfrac{V}{1 + w_f}$

$\therefore \quad$ Hull efficiency $= \dfrac{1 + w_f}{1 + a} \quad \text{and} \quad (1 + a) = \dfrac{1}{(1 - t)}$
$\phantom{\therefore \quad \text{Hull efficiency}} = (1 + w_f)(1 - t)$

Propeller Efficiency

The ratio P_T/P_D is the propeller efficiency and the ratio P_{D1}/P_D is called the relative rotative efficiency (RRE) where P_{D1} is the power to the propeller in open water; with no hull attached.

RRE = screw efficiency behind hull ÷ screw efficiency in the open.

Quasi-Propulsive Coefficient (QPC)

This term now replaces the term formerly known as the propulsive coefficient

$$\frac{P_E \text{ naked.}}{P_S}$$

$$\text{QPC} = \frac{P_E \text{ naked} + \text{allowances for appendages, weather etc.}}{P_D}$$

where the allowance is generally in the region of 8 per cent for appendages and air resistance in single screw ships, with an addition for bossings in twin screw ships, and another, say 15 per cent is added for weather resistance in average sea conditions. Also

$$\text{QPC} = \text{hull efficiency} \times \text{open efficiency of propeller} \times RRE$$
$$= (1 + w_f)(1 - t) \times \text{open efficiency of propeller} \times RRE$$

and

$$\text{hull efficiency} = \frac{\text{QPC}}{\text{screw efficiency behind hull}}$$

A formula given by Emerson which enables a close estimate to be made to the value of QPC is as follows:

$$\text{QPC} = 0.85 - \frac{N_{122}}{900}$$

and

$$N_{122} = N_D \sqrt{\frac{L_D}{122}}$$

where N_{122} = rev/min for a ship of length 122 m
N_D = rev/min for required design
L_D = length of new design.

EXAMPLE 18.6.
Estimate the QPC for a ship of $L = 142$ m; $N = 105$ rev/min.

$$\text{QPC} = 0.85 - \frac{105 \sqrt{142/122}}{900}$$
$$= 0.85 - 0.13$$
$$= 0.72$$

EXAMPLE 18.7
Determine the hull and the propeller efficiency given that $w_f = 0.24$; $t = 0.21$; QPC $= 0.67$; RRE $= 0.99$.

$$\text{QPC} = (1 + w_f)(1 - t) \times \text{open efficiency} \times \text{RRE}$$
$$0.67 = 1.24 \times 0.79 \times \text{open efficiency} \times 0.99$$
$$\text{open efficiency} = \frac{0.67}{1.24 \times 0.79 \times 0.99} = 69\%$$
$$\text{Hull efficiency} = 1.24 \times 0.79 = 98\%$$

PROPULSION AND PROPELLERS

EXAMPLE 18.8.
Given that QPC = 0·71; screw efficiency = 0·68; w_f = 0·22. Determine the thrust deduction factor.
From

$$\text{Hull efficiency} = \frac{\text{QPC}}{\text{screw eff. behind hull}}$$

$$(1 + w_f)(1 - t) = \frac{\text{QPC}}{\text{screw eff. behind hull}}$$

$$(1 + 0·22)(1 - t) = \frac{0·71}{0·68} = 1·04$$

$$1 - t = \frac{1·04}{1·22} = 0·85$$

$$t = 0·15$$

The Screw Propeller

As previously stated the propulsive device most commonly used is the screw propeller. The screw propeller is the usual method by which a thrust is developed by the propelling machinery to overcome the resistance of the ship and produce motion.

Fundamentally, the marine screw propeller may be regarded as a helicoidal surface which, on rotation, screws its way through the water, driving water aft and the ship forward.

There are various special types of propellers such as

1) Controllable pitch propeller
2) Voith-Schneider propeller
3) Shrouded propellers

but the present concern is with fixed pitch propellers.

A screw propeller has two or more fixed blades projecting from a boss. The surface of each blade when viewed from aft is called the *Face*; it is the driving surface when producing an ahead thrust. The other surfaces of the blade is called the *Back*. The *Leading Edge* of the blade is the edge which when the ship is driven ahead, first cuts the water; the other edge is termed the *Trailing Edge*.

Propellers can be designed to turn in either direction in producing an ahead thrust. If they turn clockwise when viewed from aft, they are said to be Right-Handed; if anti-clockwise, they are said to be Left-Handed. In twin screw ships the starboard propeller is normally right-handed and the port propeller left-handed. Thus the propellers are outward turning as in Figure 18-1. In this way cavitation is reduced—see later.

Other important terms are:

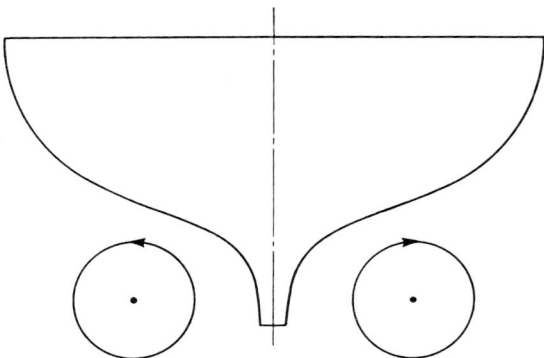

FIG. 18-1—*Twin screw ship—view from aft.*

Diameter (D) is the diameter of the circle swept out by the tips of the blades.

Pitch (P) is the distance any specified point on the face of the blade would move forward in one revolution.

$$\text{Pitch ratio} = \frac{\text{Pitch}}{\text{Diameter}} = \frac{P}{D} = a$$

Other things being equal, the thrust developed by a propeller varies directly with the surface area, excluding the boss. This area can be described in several ways. The *Developed Blade Area* is the sum of the face area of all the blades; the boss is excluded. The *Projected Area* is the projection of the blades on to a plane normal to the shaft axis. The *Disc Area* is the area enclosed within the tip circle, and is equal to $\pi D^2/4$ where D is the diameter of the propeller.

The *Blade Area Ratio* (BAR) is the ratio of the developed blade area (A_D) to the disc area.

$$BAR = \frac{4A_D}{\pi D^2}$$

Geometry of the Screw Propeller (Figure 18-2)

The faces of the blades are portions of a true helical surface, i.e. a surface swept out by a straight line AB one end of which (A) advances uniformly along the axis OO' whilst the line itself rotates about the point A with uniform angular speed. When the generating line has made a complete revolution and is in the position A'B' the distance it has advanced AA' is called the face or geometrical pitch.

Figure 18-3 shows a typical propeller design.

Slip

The distance advanced by a propeller during one revolution when

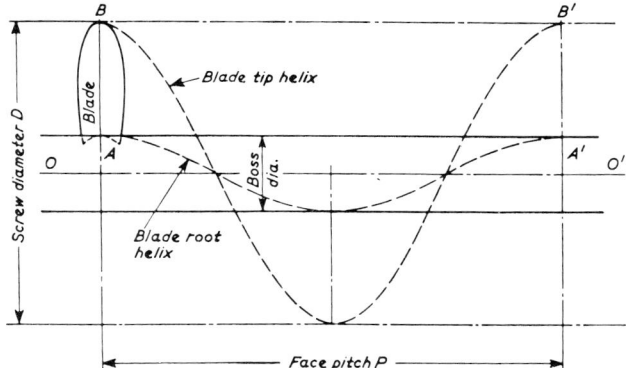

FIG. 18-2—*Geometry of screw propeller.*

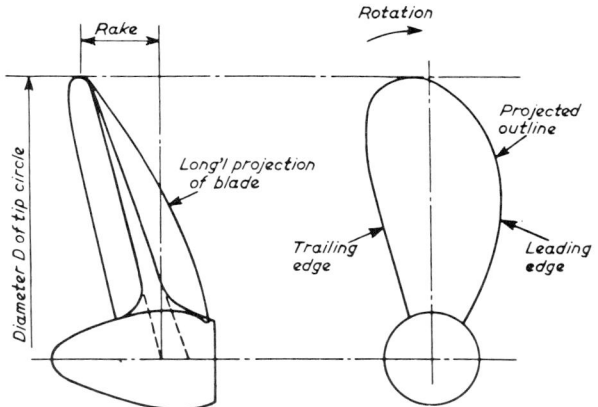

FIG. 18-3—*Typical propeller design.*

delivering no thrust is termed the analysis pitch. This is rather greater than the geometrical pitch of the propeller. When developing thrust, the propeller advance per revolution is less than the analysis pitch. The difference is termed the *Slip*.

$$\text{Slip in metres} = P - \frac{30 \cdot 864 V}{N}$$

where P = pitch in metres; V = speed in knots; N = rev/min.
Note: 1 nautical mile = 1852 m; ∴ 1 knot = 30·864 m/min

$$\text{Slip Ratio } (S) = \frac{P - (30 \cdot 86 V/N)}{P} = \frac{PN - 30 \cdot 86 V}{PN}$$

$$\therefore \quad (1-S) = \frac{30 \cdot 86 V}{PN}$$

Apparent Slip

When dealing with propellers fitted to ships as distinct from propellers tested in open water, the speed of the ship is used in the assessment of the ship and since the speed of the ship (V) differs from the speed of the water in way of the propellers (V_a).

$$\text{Apparent slip} = \frac{PN - 30 \cdot 86 V}{PN}$$

P = metres
N = rev/min
V = knots

True Slip

When dealing with model propellers run in open water, the actual speed of advance of the propeller relative to the water is known and

$$\text{True Slip} = \frac{PN - 30 \cdot 86 V_a}{PN}$$

V_a = speed of advance in knots

EXAMPLE 18.9.

A ship at a speed of 15 knots has the following particulars; $Ps = 3050$ kW; rev/min = 95; Propeller thrust = 358 kN; apparent slip = 0. For a Froude wake factor of 0·45 and a thrust deduction factor of 0·20 determine the real slip and the P_E.

$$V = V_a(1 + w_f)$$

$$V_a = \frac{15}{1 \cdot 45} = 10 \cdot 35 \text{ knots}$$

apparent slip =

$$\frac{PN - 30 \cdot 86 V}{PN} = 0$$

$PN - 30 \cdot 86 V = 0$

$$P = \frac{30 \cdot 86 \times 15}{95}$$

$= 4 \cdot 86$ m

Real slip =

$$\frac{PN - 30 \cdot 86 V_a}{PN}$$

$$= \frac{4 \cdot 86 \times 95 - 30 \cdot 86 \times 10 \cdot 35}{4 \cdot 86 \times 95}$$

$= 0 \cdot 30$

$$\frac{R}{T} = (1 - t)$$

$$\frac{R}{358} = (1 - 0 \cdot 2) = 0 \cdot 8$$

PROPULSION AND PROPELLERS 311

$\therefore$ $R = 286 \cdot 4$ kN
$P_E = R_t \times V \times 0 \cdot 5144$
$= 286 \cdot 4 \times 15 \times 0 \cdot 5144$
$= 2210$ kW

EXAMPLE 18.10.

A propeller of pitch 4·4 m has an efficiency of 65 per cent. At 116 rev/min the real slip is 0·35 and the P_D is 2750 kW. Determine the thrust of the propeller.

Propeller efficiency $= \dfrac{P_T}{P_D}$ 　　　　True slip $= \dfrac{PN - 30 \cdot 86\, V_a}{PN}$

$0 \cdot 65 = \dfrac{P_T}{2750}$ 　　　　$0 \cdot 35 = \dfrac{4 \cdot 4 \times 116 - 30 \cdot 86\, V_a}{4 \cdot 4 \times 116}$

$\therefore P_T = 1790$ 　　　　$510 - 30 \cdot 86\, V_a = 178$

$P_T = T \times V_a$ 　　　　$V_a = 10 \cdot 75$ knots

$1790 = T \times 10 \cdot 75 \times 0 \cdot 5144$
$T = 324$ kN

EXAMPLE 18.11.

A propeller has a diameter of 4·26 m with pitch ratio 1·1. At 120 rev/min the ship speed is 16 knots. Propeller efficiency is 68 per cent; Froude wake fraction is 0·22 and thrust deduction factor 0·15. For a delivered power of 1860 kW determine the P_T, P_E, QPC; the actual thrust and the true slip.

Propeller eff. $= \dfrac{P_T}{P_D} = 0 \cdot 68$ 　　　　$V_a = \dfrac{V}{1 + w_f} = \dfrac{16}{1 \cdot 22} = 13 \cdot 1$ knots

$P_T = 1860 \times 0 \cdot 68$ 　　　　$P_T = T \times V_a$
$= 1260$ 　　　　$1260 = T \times 13 \cdot 1 \times 0 \cdot 5144$
　　　　　　　　　$\therefore T = 187$ kN

$\dfrac{R}{T} = 1 - t$

$R = T(1 - t)$ 　　　　$(1 + w_f)(1 - t) = \dfrac{\text{QPC}}{\text{screw eff.}}$

$= 187 \times 0 \cdot 85$
$= 159$ kN 　　　　$1 \cdot 22 \times 0 \cdot 85 = \dfrac{\text{QPC}}{0 \cdot 68}$

$P_E = R_t \times V \times 0 \cdot 5144$
$= 159 \times 16 \times 0 \cdot 5144$ 　　　　$\therefore$ QPC $= 0 \cdot 71$
$= 1310$ kW

$\dfrac{P}{D} = 1 \cdot 1$ 　　　　True slip $= \dfrac{PN - 30 \cdot 86\, V_a}{PN}$

$\therefore$ Pitch $= 1 \cdot 1 \times 4 \cdot 26$ 　　　　$= \dfrac{4 \cdot 7 \times 120 - 30 \cdot 86 \times 13 \cdot 1}{4 \cdot 7 \times 120}$

$= 4 \cdot 7$ m 　　　　$= 0 \cdot 28$

EXAMPLE 18.12.

A propeller has a diameter of 5·2 m, a pitch ratio of 1·12 and works at an apparent slip of 7 per cent. For a ship speed of 17·5 knots determine the rev/min.

$$P/D = 1·12$$
$$\therefore P = 5·2 \times 1·12$$
$$= 5·82 \text{ m.}$$

$$\text{apparent slip} = \frac{PN - 30·86\ V}{PN}$$

$$0·07 = \frac{5·82\ N - 30·86 \times 17·5}{5·82\ N}$$

$$0·407N = 5·82N - 540$$
$$5·41N = 540$$
$$N = 100 \text{ rev/min}$$

Propeller Design

Model test data for the purpose of design are derived from systematic tests on groups of propeller models which differ, generally, in only one particular at a time. The basic variable for each group is the pitch ratio (a) where

$$a = \frac{P}{D} \qquad \begin{array}{l} P = \text{propeller pitch} \\ D = \text{propeller diameter} \end{array}$$

The first model experiments with a series of geometrically similar screws of different pitch ratios were made by R. E. Froude and the results published in 1908. Since that time many results have become available and these are, in general, presented in the form using the Taylor constants.

The Taylor constants are:

$$B_P = 0·2198 \frac{NP^{1/2}}{V_a^{2·5}}$$

$$B_U = 1·158 \frac{NU^{1/2}}{V_a^{2·5}}$$

$$\delta = 3·2808 \frac{ND}{V_a}$$

$$e = \frac{P_T}{P_D} = \text{efficiency}$$

where N = rev/min; P = delivered power in kW; V_a = speed of advance in m/sec. for B_P; V_a = speed of advance in knots for B_U and δ; U = thrust power in kW; D = diameter in metres.

EXAMPLE 18.13.

Determine the values of B_P, B_U δ and e from the following data: Propeller diameter = 5·65 m; V_a = 11·4 knots; N = 115 rev/min; P_D = 3100 kW; P_T = 2140 kW.

$$V_a = 11·4 \times 0·5144 = 5·85 \text{ m/sec.}$$

$$B_P = 0·2198 \frac{NP^{\frac{1}{2}}}{V_a^{2·5}}$$

$$= 0·2198 \frac{115 \times 3100^{\frac{1}{2}}}{5·85^{2·5}} = 17·0$$

$$B_U = 1·158 \frac{NU^{1·2}}{V_a^{2·5}}$$

$$= 1·158 \times \frac{115 \times 2140^{\frac{1}{2}}}{11·4^{2·5}} = 14·04$$

$$\delta = 3·2808 \frac{ND}{V_a}$$

$$= 3·2808 \times \frac{115 \times 5·65}{11·4} = 187$$

$$e = \frac{P_T}{P_D} = \frac{2140}{3100} = 69 \text{ per cent}$$

The results for a family of screws of the same geometrical design but differing in pitch ratio are presented in a chart such as is shown in Figure 18-4. Curves

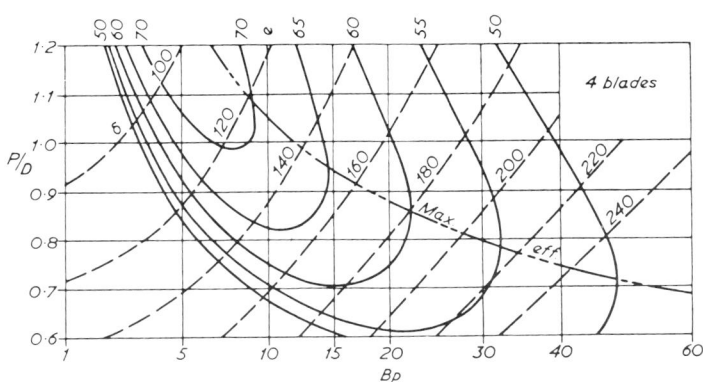

FIG. 18-4—B_p-δ Diagram.

of δ and e are plotted on a base of B_P or B_U and with an ordinate of pitch ratio (a).

The power, speed, speed of advance and revolutions can be estimated when dealing with a propeller design. From the charts the best combination of diameter, pitch and revolutions can be selected.

If an ordinate is erected on the chart at the calculated value of B_P or B_U it will be seen that there is some value of the pitch ratio (a) above or below which efficiency falls. This is the optimum pitch ratio which should be chosen. Thus the value of δ is obtained; this enables the diameter (D) to be calculated and consequently the pitch (P).

Design Procedure.

The various stages in the design procedure using the B_P δ diagrams can be summarized as follows:

1) Estimate the P_E (naked) required for the speed specified using one of the standard series applicable to the ship type under consideration.
2) Add to the P_E (naked) the allowances for appendages and weather. For appendages in single screw ships this is about 8 per cent and in twin screw about 12 per cent. For weather it is fairly normal to add a further 15 per cent. On certain routes such as North Atlantic the allowance will be much greater.
3) Obtain a value for the QPC and hence derive the P_D.
4) Determine the wake fraction from a similar ship and obtain V_a.
5) Choose a value for the rev/min. This will probably be determined by the machinery manufacturer.
6) Calculate B_P and determine from the chart the value of the optimum pitch ratio and δ.
7) Use these values to obtain the diameter and pitch.

EXAMPLE 18.14.

By the Taylor expressions and the chart of Figure 18-4 determine the optimum dimensions for a propeller in keeping with the following design conditions:

$P_D = 4800$ kW; $N = 100$ rev/min; V of ship $= 16$ knots; $w_f = 0.45$.

$$V_a = \frac{V}{1 + w_f}$$

$$= \frac{16}{1.45}$$

$$= 11.04 \text{ knots}$$

$$B_P = 0.2198 \frac{NP^{\frac{1}{2}}}{V_a^{2.5}}$$

$$= 0.2198 \times \frac{100 \times 4800^{\frac{1}{2}}}{[11.04 \times 0.5144]^{2.5}}$$

$$= 19.8$$

From chart of Figure 18-4 when $B_P = 19.8$; $\delta = 173$; $e = 0.615$; $P/D = 0.885$.

From
$$\delta = 3.2808 \frac{ND}{V_a}$$

$$D = \frac{\delta V_a}{3.2808 N} = \frac{173 \times 11.04}{3.2808 \times 100} = 5.82 \text{ m}.$$

$$a = P/D = 0.885$$
∴ $$P = 0.885 \times 5.82$$
$$= 5.15 \text{ m.}$$

Assume maximum diameter restricted to 5·0 m.

$$\delta = 3.2808 \frac{ND}{V_a} = 3.2808 \frac{100 \times 5.0}{11.04} = 149$$

From diagram
$$a = 1.162$$
∴ $$P = 1.162 \times 5 = 5.81 \text{ m.}$$
$$e = 0.585$$

Loss of efficiency = 61·5 per cent − 58·5 per cent
$$= 3\%$$

Propellers and the Law of Similarity

The use of model experiments to determine ship resistance was followed by a similar application of model technique to screw propellers. Froude's Law of Comparison may be used to predict the performance of geometrically similar propellers. The results are usually plotted in the form of non-dimensional coefficients.

Dimensional analysis has led to three basic coefficients.

Thrust coefficient $K_T = \dfrac{T}{\rho N^2 D^4}$

Torque coefficient $K_Q = \dfrac{Q}{N^2 D^5}$

Advance coefficient $J = \dfrac{V_a}{ND}$

T = Newtons; Q = Nm; N = rev/sec; D = metres; V_a = metre/sec; = 1025 kg/m³

In these the product ND is a measure of the rotative speed of the propeller. The propeller efficiency e is given by

$$e = \frac{\text{useful output}}{\text{input}}$$

$$= \frac{TV}{Q \times 2\pi N} = \frac{K}{K_Q} \times \frac{J}{2\pi}$$

If a ship and model have a ratio of dimensions equal to m, then the following relationships exist as shown in Table 18.1.

TABLE 18.1.

	For Ship Full Size Propeller	For Model Propeller
Length	$L = ml$	l
Speed	$V = m^{\frac{1}{2}}v$	v
Diameter of Propeller	$D = md$	d
Pitch of Propeller	$P = mp$	p
Revolutions of Propeller	N	$n = m^{\frac{1}{2}}N$
Slip	$S = s$	s
Thrust	$T = m^3 t$	t
Torque	$Q = m^4 q$	q
Thrust Power	$U = m^{3 \cdot 5} u$	$u = tv$
Delivered Power	$P = m^{3 \cdot 5} p$	$p = 2\pi nq$
Efficiency	$e = \dfrac{U}{P} = \dfrac{K_T}{K_Q} \times \dfrac{J}{2\pi}$	$e = \dfrac{u}{p}$
Disc Area	$A = m^2 a$	a

EXAMPLE 18.15.

A model propeller 0·41 m in diameter with a pitch ratio of 1·25 produces thrust of 266 newtons when advancing at a speed of 6·5 knots at 800 rev/min the torque being 16·4 Nm. Determine for a similar ship propeller 4·3 m in diameter the pitch, rev/min thrust, torque P_T, P_D and efficiency.

Model propeller diameter = 0·41 m $p/d = 1·25$
Ship propeller diameter = 4·3 m $\therefore p = 0·41 \times 1·25 = 0·508$ m

Ratio $m = \dfrac{4·3}{0·41} = 10·5$

For Ship

Pitch $P = mp = 10·5 \times 0·508 = 5·33$ m

rev/min $N = \dfrac{n}{m^{\frac{1}{2}}} = \dfrac{800}{\sqrt{10·5}} = 247$

Thrust $T = m^3 t = 10·5^3 \times 266 = 308000\ N = 308\ kN$

Torque $Q = m^4 q = 10·5^4 \times 16·4 = 196500\ Nm$

Speed $V = m^{\frac{1}{2}} v = \sqrt{10·5} \times 6·5 = 21$ knots.

$P_T = T \times V \times 0·5144$
$= 308 \times 21 \times 0·5144$
$= 3320$ kW

$P_D = \dfrac{2\pi QN}{1000} = 2\pi \times 196500 \times \dfrac{247}{60} \times \dfrac{1}{1000} = 5080$ kW

Efficiency $= \dfrac{P_T}{P_D} = \dfrac{3320}{5080} = 65·5$ per cent

PROPULSION AND PROPELLERS 317

EXAMPLE 18.16.
Given the following data for a propeller: Diameter = 4·88 m; K_Q = 0·015; K_T = 0·097; J = 0·633; N = 120 determine the P_D; P_T; V_a; Efficiency.

From $K_Q = \dfrac{Q}{\rho N^2 D^5}$

Torque $Q = K_Q \rho N^2 D^5$
$= 0·015 \times 1025 \times \left(\dfrac{120}{60}\right)^2 \times 4·88^5$
$= 170200\ Nm$

$P_D = \dfrac{2\pi QN}{1000}$

$= 2\pi \times 170200 \times \dfrac{120}{60} \times \dfrac{1}{1000}$

$= 2140$ kW

From $K_T = \dfrac{T}{\rho N^2 D^4}$

Thrust $T = K_T \rho N^2 D^4$
$= 0·097 \times 1025 \times \left(\dfrac{120}{60}\right)^2 \times 4·88^4$

$= 225500\ N$
$= 225·5\ KN$

$J = \dfrac{Va}{ND}$

$V_a = JND = 0·633 \times \left(\dfrac{120}{60}\right) \times 4·88$

$= 6·18$ m/sec

$P_T = T \times V_a$
$= 225·5 \times 6·18$
$= 1390$ kW

Efficiency $= \dfrac{P_T}{P_D} = \dfrac{1390}{2140} = 65$ per cent

EXAMPLE 18.17.
The undernoted data for a propeller gives values of pitch ratio and efficiency for the optimum efficiency at different values of B_P.

B_p	12	14	16	18
δ	144	154	162	170
Pitch Ratio	0·892	0·856	0·830	0·806
Efficiency	0·712	0·694	0·680	0·666

From a plot of these on a base of B_p determine the diameter, pitch and efficiency of a propeller to absorb P_D 1950 kW at 115 rev/min and a speed of advance of 11·1 knots.

$$B_p = 0.2198 \frac{NP^{\frac{1}{2}}}{V_a^{2/5}} \qquad V_a = 11 \cdot 1 \times 0.5144$$
$$= 5 \cdot 71 \text{ m/sec}$$

$$= 0.2198 \times \frac{115 \times 1950^{\frac{1}{2}}}{5 \cdot 71^{2/5}}$$

$$= 14 \cdot 33$$

From diagram at $B_p = 14 \cdot 33$. See Figure 18–5.

$$\delta = 155 = 3 \cdot 2808 \frac{ND}{V_a} = 3 \cdot 2808 \times \frac{115 \times D}{11 \cdot 1}$$

∴ Diameter = 4·55 m.

Pitch ratio $= 0 \cdot 85 = \dfrac{P}{D} = \dfrac{P}{4 \cdot 55}$ ∴ $P = 3 \cdot 86$ m

Efficiency = 69·2 per cent.

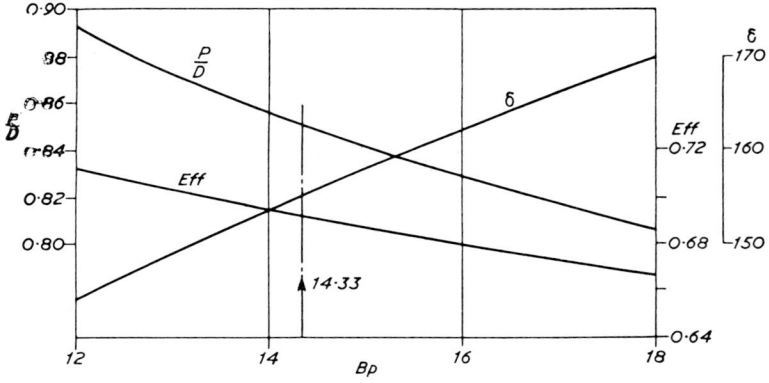

Fig. 18–5—*Example 18.17.*

Cavitation

The thrust and torque of the propeller depend upon the lift and drag characteristics of the blade sections. The lift on the section is produced partly from the suction on the back of the blade and partly from the positive pressure on the driving face. Normally the suction force is about four times as great as the pressure force.

With increase of propeller revolutions the peak of the pressure reduction curve increases and if at any point the local pressure on the blades falls below the vapour pressure a cavity will be formed filled with water vapour and air. This phenomenon leads to loss of thrust and torque, increase in the revolutions and ultimately the cavitation bubbles collapse. This can lead to intense local pressures which may cause pitting or erosion of the propeller blades. Pitting creates a weaker propeller and also increases the surface irregularities of the blades.

Singing

Before the beginning of cavitation, the blades of a propeller may give out a high-pitched note. This singing, as it is called, is due to the elastic vibration of the material excited by the resonant shedding of non-cavitating eddies from the trailing edge of the blades. This fault may be eliminated by a change in the shape of the trailing edge or increased damping of the blade.

Propeller Tests in Open Water

Probable propeller efficiency is obtained from methodical series testing of model propellers in open water. Testing of this nature eliminates the effects of cavitation and the actual flow of water into a propeller behind a particular ship form. In this way it is possible to make comparisons of different propellers on a consistent basis.

Such tests are conducted in a ship experimental tank with the propeller mounted forward of a streamlined casing which contains the drive shaft. An electric motor on the carriage drives the propeller. The thrust, torque, rev/min of propeller and the speed of the carriage are recorded and from this data K_T, K_Q, J and e can be calculated.

Cavitation Tunnel Tests

The first experiments on cavitation were recorded in 1897 by Parsons. They were part of a programme of investigation into the propulsion of the *Turbinia* which was the first stage in the introduction of the Parsons turbine as a prime mover for use in ships. The experiments were made with a small propeller operating in a closed channel fitted with glass panels. It could be seen that bubbles of air and vapour first collected near the tips of the blades at the leading edges and as the propeller speed was increased there was an extension of the region covered by the bubbles. By this the connexion between operating conditions of model and full-size propellers was approximately realized. To study cavitation it is necessary to have a closed channel or duct containing water at reduced pressure.

The adoption of such methods of testing was slow and it was not until about 1930 that the first of the modern propeller-testing tunnels was in operation. There are now several such tunnels in use. The water flows in a closed circuit and by means of an air pump the pressure on the water in the circuit can be reduced much below that due to the head plus the atmosphere. The tunnel is arranged in a vertical plane with the propeller in the upper horizontal member so that the absolute pressure at the propeller can be reduced to a minimum.

The experiments are, in general, conducted under the following conditions:

1) the water speed is made as high as possible to keep Reynold's number high and thus avoid serious scaling of skin friction
2) the model propeller has as large a diameter as is possible within the limits of the tunnel size
3) the model is run at the correct J value
4) the pressure in the tunnel is lowered to produce the correct cavitation number at the propeller axis.

Tunnels have produced a great deal of information on cavitation and the various forms it can take.

Speed Trials

Measured Mile Trials. When a ship has been completed speed trials are run to confirm that the ship can meet the contract requirements with regard to power, speed and fuel consumption. Such trials make available valuable data for design purposes.

Progressive Speed Trials. The practice is to run the ship at a series of speeds from a low speed up to the highest speed attainable. The practice of running progressive trials was introduced by W. Denny and consisted of making a series of runs over a measured mile course at a series of speeds, recording the time taken for each run and measuring and recording the power and propeller revolutions during each run.

A speed/power curve can be plotted from the records of a progressive trial and such a record is of the highest value as data for design purposes.

The trials are conducted over a measured mile course, Figure 18-6. Two pairs of posts, A B and C D, are prominently marked and set up on land one nautical mile apart and the ship's course is steered at right angles. To eliminate the effect of tide, several runs are taken both with and against the tide. To ensure accuracy in the determination of the speed of the ship from the observed times of successive runs over a measured mile course the following should be noted.

1) The successive runs should be made in opposite directions over the same course
2) Immediately on completing a run the ship should be turned off the course under a small angle of helm and run a sufficient distance before the turn so as to give a straight run prior to re-entering the mile course as shown in Figure 18-6.
3) The time intervals between the runs on the mile should be about equal and as short as possible subject to securing uniform speed on the mile.

PROPULSION AND PROPELLERS   321

4) The mean speed (V) and mean revolutions per minute should be determined by the mean of means method or from the averaging formulae, as

$$V = \tfrac{1}{8}(V_1 + 3V_2 + 3V_3 + V_4) \text{ for 4 runs}$$

or

$$V = \tfrac{1}{32}(V_1 + 5V_2 + 10V_3 + 10V_4 + 5V_5 + V_6) \text{ for 6 runs}$$

For revolutions per minute use $R_1 R_2 R_3$ etc. in lieu of $V_1 V_2 V_3$ etc.

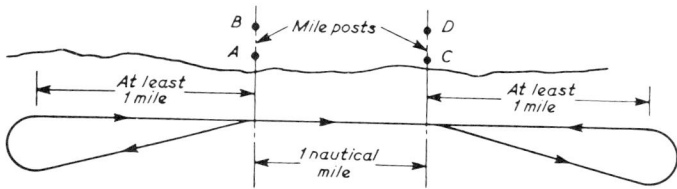

FIG. 18-6—*Measured mile course.*

EXAMPLE 18.18.

A ship on a measured mile course records speeds of 14·52; 14·92; 14·50; 14·90; 14·48; 14·88 knots for six consecutive runs at regular time intervals. Determine the mean speed.

Measured Speeds Knots	Means First	Second	Third	Fourth	Fifth
14·52					
	14·72				
14·92		14·715			
	14·71		14·710		
14·50		14·705		14·705	
	14·70		14·700		14·70
14·90		14·695		14·695	
	14·69		14·690		
14·48		14·685			
	14·68				
14·88					

Mean ship speed = 14·70 knots.
By the formula as given above for 6 runs
$V = 1/32 \,[14·52 + 5 \times 14·92 + 10 \times 14·50 + 10 \times 14·90 + 5 \times 14·48 + 14·88]$
   $= 1/32 \times 470·40$
   $= 14·70$ knots as above.

It is essential to run measured mile trials in deep water. If the water is not deep the natural stream lines are not formed round the ship and this, in general, leads to increased resistance.

The British Ship Research Association has issued a Code of Procedure for Measured Mile Trials.

INDEX

A

Abbreviations, 123, 124, 125
Activated fin, 282
Active systems, 282
Admiralty coefficient, 298
After body, 113
After perpendicular, 111
Air pipes, 58
Air resistance, 286
Aluminium alloys, 73
Amidships, 111
Angle when turning, 265
Appendage resistance, 286
Areas, 127
Area of waterplane, 121
Automatic control, 23
Automatic steering, 23

B

Baseline, 112
Beams, 69
Beaufort scale, 5
Bilge keels, 281
Block coefficient, 84, 117
BM curve, 223
BM approximation, 228
BM_L, 191
Body plan, 114
Boiler seatings, 59
Bonjean curves, 117, 206
Breadth moulded, 111
Brittle fracture, 72
Bulk carriers, 10, 32–34, 106
Bulkheads, 64
Bulkhead deck, 205
Buoyancy, 44
Buoyancy distribution, 220
Burrill, 260

C

Camber, 112
Cargo ships, 13, 16, 28
Car ferry, 27
Cathodic protection, 76
Cavitation, 318, 319

Centre girder, 56
Centre of buoyancy, 115, 151
    flotation, 115, 142
    gravity, 115
    pressure, 248
Centroids, 140
Change of draught, 159
    trim, 191, 192
Chemical carriers, 41, 105
Classification, 6, 22, 79
Coefficient, block, 117, 155
    mid-area, 117, 155
    prismatic, 117, 155
    waterplane, 117, 156
Cofferdam, 56
Containers, 36
Container ships, 17, 36, 106
Constant draught aft, 200
Controllable pitch, 307
Corresponding speed, 287, 290
Corrosion, 74
Crew accommodation, 19, 78
Cross curves, 184
Cross-channel ships, 15, 27
Curve of buoyancy, 44

D

Dangerous goods, 106
Deadweight, 18, 115
    coefficient, 115
    scale, 115
Decks, 62
Deck erections, 9
Deep tanks, 66
Definitions, 111
Deflexion, 241
Delivered power, 301
Density effect, 291
Depth moulded, 111
Denny, 275
Diesel engines, 20
Displacement, 18, 84, 121, 150
    curve, 152
    trimmed WL, 193
Direct flooding, 206
    A.P.; F.P., 211

324 INDEX

Docking, 201
   stability, 202
Double bottom, 56, 58
Draught, 7, 111
Dredgers, 15, 17
Duct keel, 54
Dynamic forces, 47

E

Eddymaking, 286
Effective power, 291, 301
Elements of structure, 53
Emerson, 306
Entrance, 114
Estimation of power, 298
Experiments on strength, 245

F

Factor of subdivision, 214
Fire, 26, 79, 105
Five-eight rule, 127, 132, 133
Fixed fins, 281
Flare, 112
Flooding, 161
Floodable length, 206
   curve, 206
Floors, 56
Food ships, 16
Fore body, 113
Forward perpendicular, 111
Form coefficients, 117, 155
   resistance, 286
Framing, 61
Free surface, 173
Freeboard, 7, 8, 19, 81, 113
   calculation, 87
Fouling, 76
Froude notation, 122
   law, 287, 289
Frequency, 258
Frictional resistance, 285
   calculation, 288
Fuel analysis, 13

G

Gas turbines, 20, 21
Gawn, 270
Girders, 69
Grain cargoes, 78
Gunwale, 63
Gyroscopes, 283
GZ, 166

H

Hatchways, 67

Heaving, 278
Heel, 112, 187
Heeling moment, 266
Higher tensile steel, 72
Hog, 44, 217
Hovercraft, 16
Hull subdivision, 25
   construction, 26
   efficiency, 302, 305
Hydrofoil, 15
Hydrostatic curves, 152

I

Icebreaker, 15
I.M.C.O., 101
Inclining experiment, 170
Indicated power, 301
Integraph, 139
Integrators, 137
Integration S.F.; B.M., 223
Intermediate ordinates, 134
International conventions, 102
   requirements, 22
Intercostal girders, 58
Internal combustion, 301
I.T.T.C., 296

J

Joessell, 270
Jumbo jet, 24

K

KB, 169
Keel, 54
Kinematic viscosity, 295

L

Laminar flow, 294
Large change in draught, 198
Law of similarity, 315
Length B.P., 111
Liddell, 261
Lifesaving appliances, 78, 103
Light, 3
   mass, 18
   and Sound signals, 78
   ship, 15, 17
   weight, 115
Lines plan, 114
Liquid gas, 15, 38–41
Lloyd's Register, 13, 22, 42, 80, 108
Load line, 8, 9, 67, 72, 82, 113
   curve, 223
Local vibration, 257

INDEX

stresses, 43, 47
Longitudinal bending, 217
  stresses, 43
Loll angle, 181

M

Machinery Certificate, 80
Marine pollution, 106
  underwriters, 22
Margin line, 26, 205
Mean ordinate, 156
  rule, 129
Merchant Shipping Act, 7, 25
Metacentric height, 10
  diagram, 168
Metacentre' transverse, 165
Measured mile, 320
M.C.T., 121, 194
Middle line plane, 113
Midship section, 111
Mild steel, 71
M.I. of waterplane, 146
Model tests, 288, 299
Modulus calculation, 229
Moment of inertia, 143
Moor, 300
Moments, 140
Moving loads, 176

N

Nuclear power, 21

O

Oil tankers, 29
Ore carriers, 10, 33
Oscillations, 277

P

Panting, 47
Parallel middle body, 113
Passenger ships, 14, 23
Passive systems, 281
Period, 4, 258
Permeability, 26, 162, 205
Pillars, 69
Pitching, 228
Planimeter, 137
Power, 122
Port authorities, 22
Powering of ships, 291
Pounding, 47
Pressure on immersed area, 246
  centre of, 248, 249, 251
Pressure, 3, 121
Propelling machinery, 20

efficiency, 302
Propulsion, 21, 301
Propeller, 122, 301, 307
  efficiency, 302, 305
  design, 312, 314
  controllable pitch, 307
  geometry, 308
  shrouded, 307
  singing, 319
Progressive speed, 320
Prefabrication, 53

Q

Q.P.C., 302, 306
Queen Elizabeth, 25

R

Radio communications, 103
Rake, 112
Rayleigh, 295
Refrigeration, 16, 35
Registration, 99
Reserve of buoyancy, 8
Resistance, 122
  residuary, 286
Removal of loads, 178
Resonant vibration, 257
Reynolds number, 294
Requirements, 18
Rise of floor, 112
Rolling, 277
Rolled sections, 50
Rudders, 121, 265
  area, 268
  bending, 273
  bow, 276
  C.P., 269
  diameter, 272
  Flettner, 276
  force, 269
  Kitchen, 276
  special, 276
  Tutin, 277
  types, 267
Run, 114
Rules for area, 127

S

Safety, 18, 102
Sag, 44, 217
Salinity, 2
Schlick, 259
Sea, 1
  chemical composition, 2
  temperature, 2

water density, 2
Section modified, 239
Seaworthiness, 7, 18
Scantlings, 7, 117
Sheer, 9, 86, 112
Ship types, 15, 23
  designer, 22
  owner, 22
  girder, 43, 47
Shipper, 22
Shaft tunnel, 59
Shell plating, 62
Ship losses, 107
Shear stress, 233
S.I. units, 119
Similar ships, 287
Simpson's rules, 127, 130–135
Six-ordinate rule, 132
Slamming, 47
Slip, 308
  apparent, 310
  true, 310
S.F. approximation, 228
  curve, 223
Stability, 7, 9, 19, 32, 104, 165, 185
Strength, 7
Statistics, 10
Steam engine, 20
  turbines, 20, 301
Statutory regulations, 77
Strength of ships, 217
Stress calculation, 229
  when inclined, 237
Stabilization, 280
Stability at large angles, 183
Suspended loads, 177
Sounding pipes, 58
Superstructures, 9, 70, 85
Survey, 78, 80
Subdivision, 104
S.O.L.A.S., 213, 215
Speed trials, 320
Swell, 4
Synchronous vibration, 257
Symbols, 120, 123–5

T

Tankers, 29, 105
Tanker tonnage, 13
Taylor, 260, 299

Tchebycheff's rule, 135
Temperature, 2
Three-ten-minus one, 143
Thrust power, 301
  deduction, 304
Tides, 3
Todd, 260
Tonnage, 78, 94
Torsionmeter, 301
Torque, 272
T.P.C., 115, 121, 153
Transverse planes, 113
  stresses, 43, 45
  stability, 165
  metacentre, 165
Trim, 112, 189
Trapezoidal rule, 128
Trawler, 15
Trochoidal theory, 5
  wave, 217
Turbulent flow, 294
Tug, 15, 17, 37
Tumble home, 112

V

Vehicular ferry, 28
Vertical prismatic, 156
Vibration of ships, 257
Virtual inertia factor, 260
Voith Schneider, 307
Volumes and centroids, 148

W

Wake, 303
  fraction, 304
Wall-sided ship, 179
Watertight bulkheads, 64, 252
  subdivision, 205
Waterplanes, 113
Waves, 4
Wave making resistance, 285
Weight curve, 221
  light, 115
Welding, 49, 50–53
Wetted surface, 121, 289
Wind, 5

Y

Yaw, 112